Sauro Succi

An Introduction to Computational Physics
Part II: Particle Methods

APPUNTI

SCUOLA NORMALE SUPERIORE
2003

ISBN: 978-88-7642-264-5

CONTENTS

VIII

PREFACE

This volume collects the second part of the Lecture Notes of the course "An Introduction to Computational Physics" held in the academic years 2000-2001 and 2001-2 for the students of the University of Pisa and Scuola Normale Superiore, at the level of third and fourth year undergraduates in physics and chemistry. This second series of lectures present material related to various types of *Particle Methods*, both deterministic and stochastic, used in modern applications of computer simulation in physics and related disciplines.

ACKNOWLEDGMENTS

I wish to express my warm gratitude to Prof. Mario Tosi, whose kind support, advice and encouragement all along, lie at the heart of this project. I would also like to thank Dr M. Trenti for his critical help with the preparation of this manuscript.

1

MOLECULAR DYNAMICS

Molecular dynamics (MD) is the sector of computational statistical mechanics dealing with the direct integration of Newtonian equations governing the microscopic dynamics of atomic and molecular constituents of matter. Originally devised to study the properties of extremely small samples of simple elements, MD plays now a major role in a vast range of disciplines, ranging from material science all the way to biology. In this chapter, we shall present the main rudiments of this important simulation technique.

1.1 Introduction

Molecular Dynamics (MD) is a simulation technique aimed at solving the Newton equations in order to compute the equilibrium and non-equilibrium properties of classical N body systems. Mathematically, MD consists in the following set of ordinary differential equations (for point-like molecules):

$$\frac{dx_i}{dt} = v_i \tag{1.1}$$

$$\frac{dp_i}{dt} = F_i(x), \tag{1.2}$$

$$i = 1, N_p$$

where $p_i = mv_i$ are the particle momenta and F_i the forces exerted upon the i-th molecule as a result of the interaction with all other molecules (x stands for the full set $x_1 \ldots x_{N_p}$ and vector notation is relaxed for simplicity). In the case of simple pairwise interactions:

$$F_i = -\nabla_i \sum_j V(x_i, x_j)$$

Upon specification of initial and boundary conditions, MD permits to obtain complete information on the dynamical evolution of the N-body system. The distinguishing feature, however, is the enormous size of the system, ideally of the order of the Avogadro's number, in practice at most a billion on present day supercomputers, and the very-long time-span of the numerical evolution. These peculiar features command special analysis.

1.2 The MD algorithm

In principle, any MD algorithm consists of two basic steps:

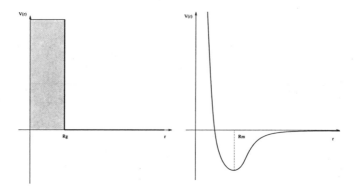

FIG. 1.1. *Typical two-body potentials: hard spheres (left) and Lennard-Jones (right). The hard speres potential represents an hard collision at the distance Rg, while nothing happens if $r > Rg$. The Lennard-Jones potential has a repulsive core ($V(r) \approx r^{-12}$ for $r \ll R_m$) and a long range interaction ($V(r) \approx r^{-6}$ for $r \gg R_m$)*

- Force Evaluation
- Particle pushing

which we now discuss in some detail.

1.3   Force evaluation

Let us refer for simplicity to pairwise central forces

$$F_i = -\nabla_i \sum_{j>i} V(r_{ij})$$

where $r_{ij} = |x_i - x_j|$ is the interparticle distance. Conceptually, the computation of F_i is very straightforward, just a double sum over all possible interacting pairs:

```
do i=1,NP-1
  f(i)=0.
  do j=i+1,NP        ! use the symmetry Fij+Fji=0
    rij=|x_i-x_j|
    f(i)=f(i)+force(rij)
    f(j)=f(j)+force(rij)
  end do
end do
```

The problem of this naive implementation is the quadratic computational complexity $C \sim N_p^2$, which makes it totally unpractical for all but the smallest simulations.

1.3.1   *Short-range potentials*

The situation is slightly better for short-range potentials (hard-spheres, Lennard-Jones):

```
do i=1,NP-1
  f(i)=0.
  do j=i+1,NP
   rij=|x_i-x_j|
   if(rij<rcut) then
    f(i)=f(i)+force(rij)
    f(j)=f(j)-force(rij)
   endif
  end do
end do
```

This is less expensive, but still requires of the order of N_p^2 cut-off tests.

1.3.2 Linked lists

An efficient way to deal with the force accumulation with linear complexity in the number of particles consists in organizing the particles according to interaction lists. First we define a control grid just for book-keeping purposes: only particles within the same grid cell interact. In other words, the mesh of the control grid is of the order of the potential range. Next, all particles belonging to the same cell are ordered along a directed list (linklist) in which each particle is given a pointer (indirect address) pointing to the next particle in the cell. The first particle is called head-of-cell (hoc) and the last particle points to null, thus marking the end of the list. As an example, for 10 particles distributed over 3 cells in 1D according the following positions:

| 3 9 5 |6   10 | 7 4 8 2 1 |

we have the following linklists (computational polymers):

```
           hoc
Cell 1:    9-->5-->3-->0
Cell 2:    10-->6-->0
Cell 3:    8-->7-->4-->2-->1-->0
```

and so on. This structure is constructed with just a few lines of code:

```
  do i=1,Ncell    !number of particle per cell init.
      nnpc(i)=0
        enddo

do p=1,NP
   i = int(x(p)/dx) + 1  ! cell index
   nnpc(i)=nnpc(i)+1      ! add a particle in a cell
   ll(nnpc(i),i)=p
enddo
```

Once the linked list is formed, only linked particles are accessed to form the overall force acting upon a given particle in the list:

```
do p1=1,NP
        f(p1)=0.
        i = int(x(p1)/dx)
c run along the linked list
        do j=1,nnpc(i)
   p2 = ll(j,i)
            if(p1.ne.p2) f(p1)=f(p1)+force(p1,p2)
        end do
        end do
```

This type of data structures are useful not only for MD simulations, but for any generic scheme dealing with disordered pairwise interactions.

1.4   Time marching

Time marching of MD equations is in principle a standard task which can be undertaken by any Ordinary Differential Equation (ODE) solver. However, the large number of degrees of freedom and time steps impose qualitative new considerations. It is worth reminding that a typical MD step is of the order of $1fs = 10^{-15}s$, so that one million steps cover only one nanosecond physical time! This is a typical example of it time-scale gap, one of the most general and compelling problems of computational statistical mechanics. High-accuracy curacy is clearly in great demand, in order to avoid dangerous accumulations of round-off errors in time.

Most popular choices are (using the following notation $h \equiv \Delta t$ and particle index relaxed for simplicity):

1.4.0.1   *Verlet*

$$x(t + h) - 2x(t) + x(t - h) = ah^2 + O(h^4) \tag{1.3}$$

$$v(t) = \frac{x(t + h) - x(t - h)}{2h} \tag{1.4}$$

Note that velocities are not needed, they are obtained by differencing the positions. This means that the velocities are only second order accurate.

1.4.0.2   *Leapfrog*

$$x(t + h) = x(t) + v(t + h/2)h \tag{1.5}$$
$$v(t + h) = v(t - h/2) + a(t)h \tag{1.6}$$

This has same accuracy for positions and velocities, but they are not defined at the same time, which is inconvenient in many respects (for instance to monitor the total energy in time).

1.4.0.3 *Modified-Verlet*

$$x(t+h) = x(t) + v(t+h/2)h \tag{1.7}$$

$$v(t+h) = v(t) + \frac{a(t) + a(t+h/2)}{2} h^2 \tag{1.8}$$

This preserves high accuracy for both positions and speeds, and keeps them synchronized on the same time sequence. An important property of time-marchers for MD is compliance with Liouville theorem, namely the volume in phase-space should be exactly conserved. Modified Verlet is the preferred choice on account of this property. Standard ODE solvers (Euler, Runge-Kutta, Gear methods) are becoming less popular because they do not fulfill this requirement.

1.4.0.4 *Sympletic time marchers*   A crucial property of very-long time integrators is to preserve the phase-space volume according to Liouville's theorem:

$$d\hat{q}d\hat{p} = dqdp,$$

where hat means $t + dt$ and (q, p) denotes the entire set of $6N_p$ phase-space coordinates.

We remind that the Liouville equation:

$$[\partial_t + p\partial_q + a\partial_p]f = 0$$

is formally solved by

$$f(t) = [e^{-(iL_q + iL_p)t}] \, f(0)$$

where $iL_q \equiv p\partial_q$ and $iL_p \equiv a\partial_p$ are the generators of translations along q and p directions of phase-space. Since these generators do not generally commute, they must be chronologically ordered. To this purpose, it is useful to use Trotter's formula. For any two generic operators A and B, Trotter's formula reads:

$$e^{A+B} = [e^{(A/N + B/N)}]^N$$

This holds exactly in the limit $N \to \infty$. For finite N, many approximate splittings are possible. With $a = A/N$, $b = B/N$, $c = [a, b]$, we have:
Simple splitting:

$$e^{a+b} = (e^a)(e^b) + O(c).$$

This is first order accurate in the commutator $c = ab - ba$, hence second order in time $(dt = t/N)$.

Strang splitting:

$$e^{a+b} = (e^{a/2})(e^b)(e^{a/2}) + O[c^2].$$

This is second order in the commutator, hence fourth-order in time.

```
        3----F
        |    |
        |    |
        |    |
I----2----1
```

Sketch 1.1: Diagrams of simple and Strang splitting
respectively. The operator a moves the
solution along the coordinate q (symbol -), whereas
operator b acts along coordinate p (symbol |).
Pathes I1F and I23F correspond to simple and Strang
splitting respectively. The terminal points I and F
stand for initial state (at time t) and final
propagated state at time t+dt.

With $a = iL_q dt$, $b = iL_p dt$, $dt = t/N$, it is easy to show that each substep
is area-preserving. Simple algebra also shows that Strang splitting is equivalent
to the Verlet algorithm. This approach is very elegant and permits to envisage
fancier area-preserving splitting schemes with higher accuracy (Trotter-Suzuki
splittings).

1.4.1   Verlet as a Trotter scheme

In this section we show that the Verlet algorithm is a special instance of the
Trotter algorithm. Consider the VV (velocity-velocity) Trotter decomposition
depicted below:

```
        3

1        2

0
---------------------
0 = (x,v,0)
3 = (x(h),v(h),h)
```

$$f_1 \equiv e^{L_p h/2} f_0$$
$$f_2 \equiv e^{L_x h} f_1$$
$$f_3 \equiv e^{L_p h/2} f_3$$

The corresponding changes in the phase space coordinates are:

1. $v_1 = v_0 + a_0 h/2$
   $x_1 = x_0$
2. $x_2 = x_0 + h v_1$
   $v_2 = v_1$
3. $v_3 = v_2 + a_2 h/2$

$$x_3 = x_2$$

Now trace this sequence of operations in reverse:

$$x_3 = x_2 = x_0 + h(v_0 + a_0 h/2) = x_0 + v_0 h + a_0 h^2/2$$

and

$$v_3 = v_2 + a_2 h/2 = v_1 + a(x_0 + hv_1)h/2 = v_0 + (a_0 + a_3)h/2,$$

which is precisely the velocity-Verlet algorithm.

1.5 Initial and Boundary Conditions

Having covered the hard-core steps of MD algorithms, we now discuss the other tasks which need to be performed in order to set up a concrete MD simulation, namely initial and boundary conditions.

1.5.1 *Initial conditions*

A common way of initializing the set of phase-space coordinates is to distribute the particle positions at random and sample the particle speeds (momenta) from a Maxwellian distribution with zero mean (no macroscopic flow) and a given temperature T_0. Actual ways to perform these tasks are discussed in the chapters devoted to Monte Carlo methods, to which the reader is directed. Here we just notice that the interparticle distance should be such as to avoid any appreciable overlap of the corresponding hard-core potentials (to prevent too strong interactions). To this purpose, particles are often initially placed on a regular lattice.

1.5.2 *Boundary conditions*

The set up of the boundary conditions depends of course on the the type of physical situation to be investigated. For the simplest case of a homogeneus system away from solid walls, periodic boundary conditions are appropriate. These consist of simply winding up particles whose positions exceed the domain size. For the simple case of a cubic box of linear size L, if it is found that $x < 0$ or $x > L$:

$$x = x - sign(x)L$$

If the system is confined within solid walls, the appropriate potential should be used to describe the microscopic interaction with the atoms in the solid wall.

```
-----------------------
0  0  0  0  0  0  0  0    atoms in the solid wall
-----------------------

x    x      x     x
                     x
      x       x
x                 x
   x       x           atoms in the internal domain
            x   x
     x        x
```

```
x       x          x      x
-------------------------
0  0  0  0  0  0  0  0    atoms in the solid wall
-------------------------
```

Sketch 2.1: Boundary conditions at solid walls
are implemented by using specific interaction
potentials between bulk atoms and atoms in the
solid walls.

1.6 Macroscopic observables

Once the state of the system is known at each given time throught the phase-space coordinates $[x_i(t), p_i(t)]$, $i = 1, N_P$, macroscopic observables are obtained by accumulation of the microscopic information. The main macroscopic observables are defined as follows:

- Number: $N(x,t) = \sum_{i \in B(x)} \delta[x_i(t) - x])$
- Speed: $u(x,t) = \sum_{i \in B(x)} v_i \delta[x_i(t) - x]/N(x,t)$
- Kinetic energy: $KE(x,t) = \sum_{i \in B(x)} \frac{mv_i^2}{2} \delta[x_i(t) - x]/N(x,t)$
- Potential energy: $PE(x,t) = \sum_{i \in B(x), j > i} V(x_i, x_j) \delta[x_i(t) - x]/N(x,t)$

where $B(x)$ denotes a box of volume $V(B)$ centered around position x. Temperature is then available as:

$$k_B T(x,t) = KE(x,t) - \frac{mu^2}{2}$$

that is, the kinetic energy in excess of mean fluid motion.

1.7 Constrained MD

MD simulations conserve the total (kinetic+potential) energy:

$$E = \frac{1}{2} \sum_i mv_i^2 + \sum_i \sum_{j>i} V_{ij} \equiv KE(t) + PE(t)$$

However, macroscopic experiments take place in different conditions, such as constant temperature, pressure or volume. As a result, useful MD simulations must take place in the corresponding NPT,NVT,.. statistical ensembles. Mathematically, this amounts to constrain the Newtonian trajectories to the corresponding hypersurface in phase-space. Imagine we wish to perform constant-temperature simulation. Since total energy is conserved, KE and PE both fluctuate in time, their sum being constant. A simple way of ensuring a constant kinetic energy is to rescale all molecular velocities by a normalization factor:

$$v_i' = v_i \sqrt{T_0/T(t)},$$

where $T(t)$ is the actual temperature value and T_0 the desired one. This rescaling, sometimes referred to as isokinetic ensemble, keeps temperature to the desired value but it does not correspond to the canonical ensemble because it fails to reproduce thermal fluctuations. A more effective and general solution to this problem consists in augmenting Newtonian equations with *thermostatted* variables, whose dynamics is designed in such a way as to keep the system trajectory around the isothermal hypersurface and allowing thermal fluctuations as well. Essentially, one adds a time-dependent control variable in the form of a friction coefficient η, which pulls the system back to the hypersurface $T = T_0$ whenever the trajectory departs from this hypersurface:

$$\frac{dx_i}{dt} = v_i \tag{1.9}$$

$$\frac{dv_i}{dt} = F_i/m - \eta v_i \tag{1.10}$$

$$\frac{d\eta}{dt} = (T - T_0)/\tau \tag{1.11}$$

From these equations, it is seen that whenever $T > T_0$, the 'pull' η grows up, thereby reducing the value of v, and viceversa. This is an example of non-Newtonian Molecular Dynamics, in that it tolerates violations of the Liouville theorem. Note that $\eta \neq 0$ corresponds to introducing compressibility effects in phase-space dynamics (expansion/contraction for $\eta > 0$ or $\eta < 0$ respectively).

1.7.1 *Holonomic constraints*

Molecular dynamics is the chief simulation tool for biological applications (for instance, organic polymers). When dealing with complex molecules, like biopolymers, many further constraints need to be accounted for, corresponding, say, to the bond lengths between the monomers along the polymer chain, their torsional degrees of freedom and so on. A typical holonomic constraint, fixed bond-length between monomers i and j, reads as follows:

$$r_{ij} = const$$

This corresponds to a set of $N(N-1)/2$ holonomic constraints on the Newtonian dynamics, N being the length of the polymer chain. A very effective algorithm to fulfill these constraints (SHAKE) is due to Ryckaert-Ciccotti-Berensen. It works on the idea of letting first the system evolve along its unperturbed (unconstrained) trajectory. A pull-back force is then computed based on the departure of the unpertubed trajectory from the desired hypersurface. For full details see the original paper cited in the reference list.

1.8 References

1. D. Frenkel, B. Smit, Understanding molecular simulation, North-Holland, 1996.
2. M. Allen, D. Tildesley, Computer simulation of liquids, Oxford Science Publications, 1987.
3. J. Ryckaert, G. Ciccotti, H. Berendsen, J. Comp. Phys. 23, 327, 1977.

1.9   Projects

Write a MD code for Lennard-Jones Argon molecules and compute the equation of state of Argon.

2

THE PARTICLE-IN-CELL METHOD

Particle-Particle (PP) methods, such as Molecular Dynamics, are very accurate because they take into explicit account all the intermolecular interactions via the the pairwise potential V_{ij}. On the other hand, the very same point makes PP methods very expensive, with a quadratic scaling of computational cost with the number of particles. In this Chapter, we shall discuss Particle-Mesh methods (PM) an efficient alternative to PP methods for systems with smooth interactions.

2.1 Introduction

Molecular interactions often split into a short-range and long-range components, corresponding to hard-core collisions and soft-core mean-field interactions. Particle-Mesh methods (PM) are designed to deal with systems where the latter interactions prevail. Examples abound in many areas of physics/engineering/biology, i.e.: plasmas, colloidal flows, quantum gases and many others.

2.2 Particle-Mesh representation

The basic idea behind PM methods is that particle interactions take place through the intermediate of a 'smooth' field varying on a space-time scale much longer than the mean interparticle separation:

$$d << s$$

where

$$d = n^{-1/3}$$

is the mean intermolecular separation, and s is the typical range of the interaction. Under these conditions, the interaction field can be placed on a lattice at a lower resolution than the interparticle mean distance. Consequently, under the assumption:

$$N_g << N_p$$

where N_g and N_p are the number of particles and grid points respectively, the computational cost of the Mesh-Particle simulation:

$$C_{PM} \sim a N_p N_g$$

can be made much smaller than the corresponding cost for a PP simulation,

$$C_{PP} \sim b N_p^2.$$

This is the central point of PM methods.

Of course, statistical accuracy requires a sufficient number of particles per grid point, typically

$$n_{pg} \equiv \frac{N_p}{N_g} \simeq 10 - 50$$

to ensure a statistical error below, say, ten percent.

2.2.1 Particle and field representation

Let us refer for concreteness to a Vlasov-Poisson system of the form:

$$\partial_t f + v \partial_x f + a \partial_v f = C(f, f) \qquad (2.1)$$

$$ma = -\partial_x \Phi \qquad (2.2)$$

$$\Delta \Phi = q \int f dv \qquad (2.3)$$

where $f(x, v, t)$ is the particle distribution function, m the particle mass and Φ the self-consistent potential obeying the Poisson equation. Finally $C(f, f)$ represents hard-core collisions, which shall be set to zero in the following (Very Dilute Gas Approximation). The distribution function at time t is represented by a set of pointlike particles at position $x_i(t)$ and speed $v_i(t)$, $i = 1, N_P$ (Klimontovich representation).

$$\tilde{f}(x, v, t) = \sum_i \delta[x - x_i(t)] \delta[v_i - v_i(t)] \qquad (2.4)$$

Here tilde enphasizes the fact that this representation inherits all statistical fluctuations associated with the discrete particles. Actual observables are derived from a corresponding ensemble-averaged distribution $f = < \tilde{f} >$, where brackets denote average over space, time and/or initial conditions.

Therefore we shall place the potential on a discrete grid $x_g = gh$, $g = 1, N_G$, with a mesh spacing $h < l_c$:

$$\Phi(x, t) = \sum_g \Phi_g W(x - x_g) \qquad (2.5)$$

where $W(.)$ are suitable shape functions centered about the spatial location x_g

FIG. 2.1. *One dimensional grid: note that the grid dimension h is greater than the mean interparticle distance. The potential field (upper curve) is slow-varying over a grid cell.*

The field $\Phi(x)$ is assumed to vary on a lengthscale l_c (coherence length) much longer than the interparticle separation d, n being the particle number density. The appropriate spatial ordering for PM methods to apply is therefore:

$$d < h < l_c$$

The $d < h$ condition ensures computational savings over a direct PP representation, whereas the $h < l_c$ inequality secures that the interactions are well resolved by the grid representation.

2.3 Computational scheme

A typical PM algorithm proceeds through the following four basic steps:

1. *Charge-assign (Particle-to-Grid)*
2. *Solve Poisson*
3. *Force assign (Grid-to-particles)*
4. *Push particles*

which we now illustrate in some detail.

2.3.1 *Charge-assign*

The task of the charge-assign step is to distribute the charge from particle to grid locations (Scatter operation):

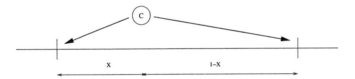

FIG. 2.2. *Charge assign: from particles to grid points*

Mathematically:

$$Q_g = \sum_i K[x_g - x_i]q_i$$

where the charge q_i stands for any observable quantity attached to the i-th particle ($n_i = 1$ for particle number). The above expression is a discrete version of the general filtering transform:

$$Q(x) = \int K(x - x'; w)q(x')dx'$$

where K is a suitable integral kernel with the properties

$$\lim_{w \to 0} K(x - x'; w) = \delta(x - x') \qquad (2.6)$$

$$\int K(x - x'; w)dx' = 1 \qquad (2.7)$$

where w is the typical width of the kernel. Clearly, the kernel represents a finite-size (dressed) particle. It is clear that each kernel gives rise to its own assignement scheme. Most practical ones are:

Nearest grid point (NGP)

$$K(x - x') = 1; \quad |x - x'| < h$$
$$K(x - x') = 0; \quad elsewhere$$

In practice, in the NGP scheme, the entire particle charge is assigned to the nearest grid point, $Q_g = q_i$, where x_g denotes the nearest grid-point to x_i. [1]

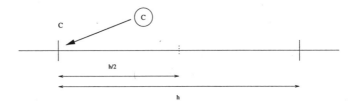

FIG. 2.3. *Nearest grid point: the charge of a particle is assigned entirely to the nearest grid point.*

Cloud-in-Cell (CIC)
The NGP kernel is obviously very crude (piecewise local interpolation) and a more accurate choice is a piecewise linear kernel centered about the particle position:

$$K(x - x') = (x - x')/h; x' - h < x < x'$$
$$K(x - x') = 1 - (x - x')/h; x' < x < x' + h$$

(note that this kernel has support of size $2h$, namely twice the support of the NGP kernel).

In practice, one assigns a fraction $K_g = 1 - h_1/h$ to the right neighbor g and $K_{g-1} = 1 - h_2/h$ to the left neighbor $g-1$. Here $h_1 = x_g - x_i$ and $h_2 = x_i - x_{g-1}$ are the right and left nearest neighbors respectively.

It is easily understood that higher order kernels provide better accuracy. This has to be balanced against the increasing computational costs.

---

[1]On a regular mesh this is easily found via a nearest integer operation (nint) as: $g(i) = nint(x_i/h)$. On irregular meshes the search is generally more expensive.

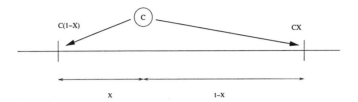

FIG. 2.4. *Cloud in cell: the charge of a particle is assigned to the two nearest grid points, linearly with the inverse of the distance*

2.3.2 *Solve the field equation (Poisson)*

Once the charge is assigned to the grid points, we can solve the field equation by any of the methods discussed in Volume I of these lectures. For the case of Poisson equation on regular grid, the spectral method is the best option. Further customized methods are discussed in the book by Hockney-Eastwood.

2.3.3 *Force-to-Particles (Force-Assign)*

Once the field is know at a subsequent time $t + dt$ by solving the field equations, we must redistribute it back to the particles (Gather operation) and compute the forces to push them forward to the next time step. This operation is dual to charge assign and amounts to a standard interpolation:

$$F(x_i) = \sum_g W[x_i - x_g]F(x_g)$$

where $F(x_g)$ is obtained from Φ_g by some standard finite-differencing, say $F(x_g) = \frac{\Phi_{g+1} - \Phi_{g-1}}{2h}$.

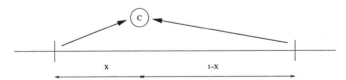

FIG. 2.5. *Force assign: from grid to particles.*

Since force-assign is dual to charge-assign, a reasonable choice is to choose the same kernel for both operations, $W = K$. However, one may choose $W \neq K$, in which case the *force-assign kernel W must be of lower or equal order than charge-assign kernel K*. Otherwise, spurious self-forces arise: the force generated by a given charge reacts back on the charge itself. (As a useful exercise, compute the force exerted upon a particle by its own field). The rationale is that the field obeys a second order equation involving the Laplacian $\Delta\Phi$ whereas the calculation of the force involves only first order derivatives, $F = -\nabla\Phi$.

2.3.4 Particle-Pushing

Once the forces on the particles are known, the particle speeds and positions can be advanced in time with any of the many time-marching schemes available from ordinary differential equation solvers or Molecular Dynamics. As discussed in Chapter I, most popular choices are ($h \equiv \Delta t$):

2.3.4.1 Verlet

$$x(t + h) - 2x(t) + x(t - h) = ah^2 + O(h^4) \tag{2.8}$$
$$v(t) = \frac{x(t + h) - x(t - h)}{2h} \tag{2.9}$$

Note that velocities are not needed, they are obtained by differencing the positions. This means that the velocities are only second order accurate.

2.3.4.2 Leapfrog

$$x(t + h) = x(t) + v(t + h/2)h \tag{2.10}$$
$$v(t + h) = v(t - h/2) + a(t)h \tag{2.11}$$

This has same accuracy for positions and velocities, but they are not defined at the same time, which is inconvenient.

2.3.4.3 Modified-Verlet

$$x(t + h) = x(t) + v(t + h/2)h \tag{2.12}$$
$$v(t + h) = v(t) + \frac{a(t) + a(t + h/2)}{2}h^2 \tag{2.13}$$

This preserves high accuracy for both positions and speeds, and keeps them synchronized on the same time sequence. An important property of time-marchers for particle methods is compliance with Liouville theorem, namely the volume in phase-space should be exactly conserved. Modified Verlet is the preferred choice on account of this property.

2.4 Convergence and numerical conservation

In an ideal simulation, the grid should have no effect on physical observables. The goal is to erase the lattice, in the sense that the discrete propagators (Green functions) should be lattice independent. In practice, this can only happen to a finite order of accuracy.

Let us define:

1. $K(x - x_g)$: Fraction of charge assigned to grid point x_g from a particle in x,

2. $G(x_g - y)$: Potential on a particle in y, due to a unit charge in x_g

By definition, the (lattice) potential on the target particle in y due to the source particle in x is given by:

$$G(x, y; N_g) = \sum_g K(x - x_g)G(x_g - y) \qquad (2.14)$$

In the continuum, G should depend only on $r = |y - x|$. The identity $x_g - y = x - y + x_g - x$ suggests an expansion of the discrete Green function in powers of the particle off-set:

$$\xi_g \equiv x_g - x,$$

This yields:

$$G(x, y; N_g) = \sum_g K(x - x_g)[G(x - y) + \xi_g G'(x - y) + \frac{\xi_g^2}{2}G''(x - y) + \ldots \quad (2.15)$$

Lattice independence requires:

$$\sum_g K(\xi_g) = C_0 \qquad (2.16)$$

$$\sum_g K(\xi_g)\xi_g = C_1 \qquad (2.17)$$

$$\sum_g K(\xi_g)\xi_g^2 = C_2 \qquad (2.18)$$

$$\ldots \qquad (2.19)$$

where parity, $K(\xi_g) = K(-\xi_g)$, has been assumed.

The quantities C_n must be lattice-free constants. In fact, apart from the case $C_0 = 1$, all other constants should be exactly zero. This is obviously guaranteed by grid-bound particles ($x = x_g$) but it cannot be true in the general case. As a result, one must accept that only a finite number of constants can be made identically zero by a proper choice of the charge and force assign schemes. The higher order terms left with $C_n \neq 0$ are clearly producing spurious forces in the equation of motion of the generic test particle:

$$m\frac{d^2y}{dt^2} = -\nabla_y \sum_x G(x, y; N_g)$$

These spurious forces are responsible for loss of spatial accuracy of the method.

Coming back to the lattice-freedom constraints (2.16), we note that the first equation can be solved with just one single grid point, by simply taking:

$$K_1 = 1.$$

The NGP assignement scheme is easily recognized.

To fulfill the first two equations, we need charge-assign on two grid-points:

$$K_1 + K_2 = 1 \qquad (2.20)$$

$$K_1 x_1 + K_2 x_2 = x \qquad (2.21)$$

which defines the CIC scheme. Higher order interpolants further improve the accuracy, but the scheme becomes correspondingly more non-local and computationally demanding. Full details on the conservation properties of the various schemes can be found in the classical book by Eastwood-Hockney.

2.5 P3M Methods

PM methods are excellent so long as the fields remain slowly-varying as compared to the interparticle mean-distance, d, namely when the potential develops no sharp features (shocks, interfaces, solitons and the like). Whenever this assumption is broken, the PM method must be combined with direct PP techniques to accomodate the short-scale physics. The result are the so called P3M (Particle-Particle+Particle-Mesh) algorithms. Briefly, in P3M, particles within a given cutoff interact pairwise, like in Molecular Dynamics, whereas particles above this cutoff are still handled by PM methods.

2.6 References

J. Eastwood-R. Hockney, Computer simulation using particles, Adam Hilger, 1976

2.7 Projects

Solve the Vlasov-Poisson equation for a one-dimensional electron plasma using the sample code given below.

2.8 Warm-up code

```
c =================================================
c particle in cell: Vlasov-Poisson
c beware: not fully tested!
c =================================================
      parameter (NG=128,NPcell=10,NP=NG*NPcell)
      dimension xp(NP),xpnew(NP),xpold(NP),forcep(NP)
      dimension xg(NG),rho(NG),phi(NG),forceg(NG)
      integer g
      real Lx
c ----------------------------------------
      NT   = 1000
      iout = 30
      iseed = 976851
      Rep=1./float(NP)
c grid
```

```
      Lx = 1.
      dx = 1./float(NG)
      dxi= 1./dx
      dt = 0.01

      xg(1) = 0.
      rho(1)= 0.
      do g   = 1,NG-1
       xg(g+1)=xg(g)+dx
      end do
c particle initial positions
      do i=1,NP
       r=ranpang(iseed)      !random number generator
       xp(i)=0.5*Lx+(r-1)*Lx*0.25
       xpnew(i)=xp(i)
       xpold(i)=xp(i)
       write(iout,*) i,xp(i),r
      end do
      pause
c -------------------------------------->
      do it=1,NT
c 1. charge assign: particles to grid
      do g=1,NG
       rho(g)=0.
      end do
      do i=1,NP
       g  =1+int(xp(i)/dx)  ! NGP
       rho(g)=rho(g)+Rep
      end do
c 2. solve the field equation

      call poisson(rho,phi,forceg,dxi)

      do i=1,NP
       g = 1+int(xp(i)/dx)
       a = dxi*(xp(i)-xg(g))

c 3. force on particles: grid to particles
      forcep(i)=forceg(g)*(1-a)+a*forceg(g+1)

c 4. particle push: simple Verlet
      xpn=2*xp(i)-xpold(i)+forcep(i)*dt*dt

c periodicity
```

```
      if(xpn.gt.LX) xpn=xpn-LX
      if(xpn.lt.0)  xpn=xpn+LX
      xpnew(i) = xpn
      end do
c prepare next step
      do i=1,NP
      xpold(i)= xp(i)
      xp(i)   = xpnew(i)
      end do
c --------------------------
      if(mod(it,200).eq.0) then
      iout=iout+1
      do g=1,NG
       write(iout,*) g,rho(g),phi(g),forceg(g)
      end do
      write(iout,'(bn)')
      endif
c -------------------------
      end do

      stop
      end
c ===================================
      subroutine poisson(rho,phi,forceg,dxi)
c ===================================
c solve tridiagonal A*phi=rho using Thomas algo
c phi(1)=phi(NG)=0
      parameter (NG=128)
      dimension rho(NG),phi(NG),forceg(NG)
      dimension A(NG),B(NG)
c ---------------------------------------------------
      phi0=0.
css bward sweep
      A(NG)=0.
      B(NG)=phi0
      al = 1.*dxi
      ad =-2.*dxi
      ar = 1.*dxi
      do g=NG,2,-1
       A(g-1)=-al/(ar*A(g)+ad)
       B(g-1)=(rho(g)-ar*B(g))/(ar*A(g)+ad)
      end do
c fward sweep
      phi(1)=phi0
```

```
do g=1,NG-2
 phi(g+1)=A(g)*phi(g)+B(g)
end do
phi(NG)=phi0

forceg(1)=-phi(1)*dxi
do g=2,NG-1
 forceg(g)=-0.5*dxi*(phi(g+1)-phi(g-1))
end do
forceg(NG)=-phi(NG)*dxi

return
end
```

3

N-BODY PROBLEMS

Many physical systems are governed by non-local, long-range interactions. Ex-amples in point are classical gravitating systems in astrophysics, one-component plasmas in condensed matter, charged colloids in physical chemistry, and many others. Long-range interactions set a significant challenge to statistical mechanics in general, and, as a consequence, to computational physics as well. Long-range interactions are computationally very demanding because they imply all-to-all interaction, namely a quadratic complexity with the number of particles. Tech-niques to bring down such quadratic complexity to a quasi-linear one are therefore greatly demanded. In this chapter we shall present a cursory view of the major ideas behind this task.

3.1 Long and short-range interactions

Let us consider a set of N particles subject to classical Newtonian gravity. The equation of motion for such a gravitational N-body systems read as follows:

$$\frac{d^2\vec{x}_i}{dt^2} = -G\vec{\nabla}_i \sum_j \frac{m_j}{|\vec{x}_i - \vec{x}_j|} \tag{3.1}$$

where G is the gravitational constant. From these equations, we observe that close encounters give rise to intense short-range interactions, whereas distant particles experience only soft, long-range, interactions. An exact computation accounting for both classes of interactions must necessarily involve N^2 complex-ity, the so called Direct Simulation approach. This approach is necessarily limited to small (order of thousands) number of bodies, and consequently new strate-gies are needed to go beyond such limitation. As usual, one tries to trade small errors for great pay-offs in computational simplicity. The main idea is to com-pute short-range interactions directly (exactly), whereas long-range interactions shall be described only through (plausible) approximations, typically multipole expansions. Hence, in the first place, we need a separation criterion to split each given configuration into a short-range (SR) and long-range (LR) partitions. In addition, since these systems are known to produce sharp density contrasts and particle agglomeration, the need arises of dynamically adaptive methods, able to adjust to the fast changes of the particle configuration. Both requisites are best met by dynamically adaptive hierarchical methods, tree-methods for short.

3.2 Basic tree algorithm

The basic idea of tree methods is to start from a reference grid and break-up it recursively until each cell contains either just one particle or no particle at all. The terminology is:

$$n = 0 \quad dead\ cell$$
$$n = 1 \quad leaf$$
$$n > 1 \quad twig$$

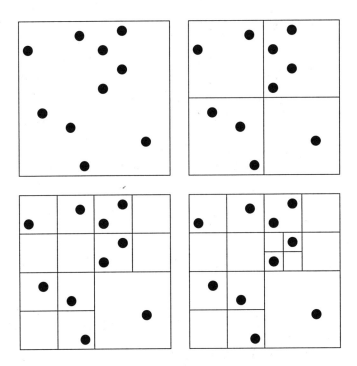

FIG. 3.1. *Tree building with a hierarchical partition method*

The N-body algorithm consists of the following three main steps:

1. *Tree construction*
2. *Force evaluation*
3. *Boundary conditions*

which we now discuss in some detail.

3.2.1 *Tree construction*

The data structure associated with the tree is constructed as follows:
Number the cells along the twig until a leaf is reached;
Then move to the next twig.

With reference to the following example:

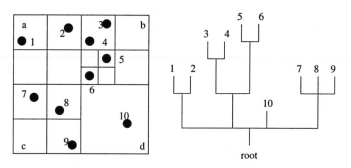

FIG. 3.2. *Example of a tree structure associated with the particle configuration.*

Recursive break-up of the cells, proceeding upwards from the root cell, generates the following data structure:

pointer	level	node	parent	1stdau	ndau	plabel
1	0	0	0	2	4	-
2	1	-1	0	6	2	-
3	1	-2	0	8	2	-
4	1	-3	0	15	3	-
5	1	1	0	-	-	10
6	2	2	-1	-	-	1
7	2	3	-1	-	-	2
8	2	-4	2	10	2	-
9	2	-5	-2	12	1	-
10	3	4	-4	-	-	3
11	3	5	-4	-	-	4
12	3	-6	-5	-	-	-
13	4	6	-6	-	-	6
14	4	7	-6	-	-	5
15	2	8	-3	-	-	7
16	2	9	-3	-	-	8
17	2	10	-3	-	-	9

Legend:

```
pointer: pointer to the parent cell
level  : level of the tree the node belongs to
node   : label of the node (negative for twigs, positive for leaves)
parent : label of the parent node
1stdau : label of the first daughter
ndau   : number of daughters
plabel : label of the particle
```

We end up with a tree made up of 10 leaves and 7 twigs (including the root). With the tree structure in place, we can proceed to the force evaluation phase.

3.2.2 Force evaluation

As already mentioned, forces must first be classified as short and long-range (SR,LR) respectively. Thus, what we need is a separation criterion.

3.2.2.1 The separation criterion The goal is that only close particles should interact pairwise exactly, whereas distant particles interactions can be dealt with approximately via the intermediate of a multipole field. More precisely: A test particle in x interacts pairwise exactly with a source particle in y only if:

$$d/s \gg 1$$

where $d = |y - x|$ is the interparticle separation and s is the size of the cluster the source particle belongs to. Otherwise, the test particle only sees average information through particle-quasiparticle interactions, as expressed by a sequence of *multipole* (MP) fields. On the tree, the following actions are taken:

1. If $s/d < \theta$:
   Particle is far, use MP interaction
2. Else
   Do until $s/d < \theta$ or a leaf is reached:
   $s/d > \theta$: Particle is close: break up the cell
   End do
   Endif

Here θ is a *tolerance* parameter which reflects the stipulated notion of closeness. The value $\theta = 0$ implies zero tolerance, which means that all interactions are treated exactly (Direct Simulation limit), whereas large θ's imply that a large fraction of interactions are treated approximately. Manifestly, θ controls the tradeoff between accuracy and computational efficiency.

theta	interaction	complexity	
0	exact	N*N	(Direct Simulation)

```
large          approx              NlogN

infinity       very approx         N       (Particle-Mesh)
```

It should be noted that multipole particle-quasiparticle interactions do *not* conserve energy exactly. To see how this happen, let us discuss the multipole interactions in some more detail.

3.2.3   *Multipole expansions*

Consider a source particle at location x_i belonging to a cluster whose center mass is in X. By letting $x_i = X + s_i$, we can use the local particle coordinate, s_i, form an expansion of the force exerted by the source particle upon a test particle in y, outside the cluster:

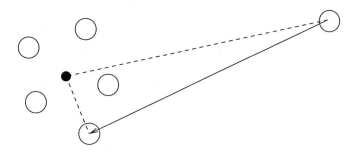

FIG. 3.3. *Force exerted from a particle inside a cluster to a particle outside the cluster. The black central point in the cluster represents the center of mass.*

$$F_i \equiv F(X + s_i) = F(x) + s_i F'(X) + \frac{1}{2} s_i s_i F''(X) + \dots \qquad (3.2)$$

Upon summing over all source particles in the cluster, we obtain:

$$F = \sum_i F_i = M f(X) + D \nabla f(X) + \frac{Q}{2} \nabla \nabla f + \dots \qquad (3.3)$$

where we have set $F_i = m_i f_i$, and

$$M = \sum_i m_i, \; D = \sum_i m_i s_i, \; Q = \sum_i m_i s_i s_i \qquad (3.4)$$

are the zeroth, first and second order multipole coefficients, namely total mass of the particles in the cluster, dipole and quadrupole moments. The multipole moments contain information on the spatial distribution of the particles in the cluster (in general D is a vector and Q a second-order tensor, but we shall stick to a one-dimensional scalar notation for simplicity). With N particles in the cluster,

N multipole moments would reconstruct this distribution exactly. The idea is to achieve the desired level of accuracy with a much smaller number of multipoles than particles in the cluster, $M << N$. Clearly, the closer x is to y, the higher the number of multipole terms to be retained in the expansion. Typically, a ratio $M/N \sim 10 - 100$, namely a few tens of MP fields versus hundreds or thousands particles in the cluster. The key to turn around the quadratic complexity is to use parent level information to compute the daughter level MP expansions, so that when moving upwards along the tree we do not need to recompute the MP expansions each time from scratch. This is made possible by the existence of the so-called shift theorems.

3.2.4 Shift theorems

Consider a mother cluster with three daughters:

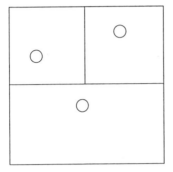

FIG. 3.4. *Mother cell with three daughters.*

Our task is to compute the multipoles of the mother cluster assuming the multipoles associated with each of the daughter cells are known. The zeroth order monopole is trivial:

$$M_0 = M_1 + M_2 + M_3$$

The dipole calculation proceeds as follows. The dipole associated to the mother cluster is

$$D_0 = \sum_i m_i(x_i - X_0)$$

where X_0 is centermass coordinate of the mother. This also rewrites as:

$$D_0 = \sum_i m_i\left[(x_i - X_k) + (X_k - X_0)\right]$$

where X_k, $k = 1,3$ are the centermass coordinates of the k-th daughter. As a result, we obtain:

$$D_0 = D_k + M_k(X_k - X_0) \equiv D_k + M_k S_k \qquad (3.5)$$

where $M_k = \sum_{i=1}^{N_k}$ is the total mass of the k-th daughter and $S_k = X_k - X_0$ is the corresponding *shift*. A similar calculation for the quadrupole yields:

$$Q_0 = Q_k + 2S_k D_k + M_k S_k S_k \tag{3.6}$$

The expressions (3.5) and (3.6) solve the problem of computing the mother MP's as a function of the daughter's ones. All higher order MP's can be computed this way. It is therefore clear that by walking top-down along the tree, *all* MP's at each level can be computed all the way down to the root. This is specular to the procedure of tree construction which proceeds bottom-up instead.

```
2:   Z    Z  Z    Z

1:      Y       Y

0:         X


Tree construction: Bottom-up (X-Y-Z)
Force evaluation : Top Down   (Z-Y-X)
```

3.3   Time marching

Once the forces are available, we need to advance positions in time. One can use Verlet-like integrators, with the additional caveat that since many interactions are computed approximately, energy is not exactly conserved and therefore accurate time integration is required. Roughly speaking, the time-step is restricted by stability conditions of the form:

$$\frac{F\delta t^2}{2m} < d_{min}$$

where d_{min} is the minimal interparticle separation. Since gravitation is attractive, particles can come very close to each other. It is clear that if particles come too close, the left hand side blows up due to $1/r^2$ factor in the force, while the right-hand-side approaches zero. This leads to a collapse of the time marcher. To prevent this catastrophic event, the gravitational interaction is smoothed out by a numerical cutoff, a, $1/r^2 \to 1/(r^2 + a^2)$. It should be noted that this breaks energy conservation (the virial theorem), and consequently the spurious effects introduced by the numerical cut-off must be monitored very carefully.

Since gravitational systems give rise to highly irregular mass distributions, it is clear that the time-step limitations may change substantially from place to place. This is why is is a common practice to advance each cluster with its own locally adaptive time-step.

3.4   Boundary conditions

Since we deal with long-range interactions, we shall consider infinite systems with periodic boundary conditions. There are essentially three methods to implement periodic boundary conditions:

1. Recirculating particles
2. Minimum image method
3. Ewald summation

3.4.1 *Recirculating particles*

This is very simple: whenever a particle quits the simulation box, it reenters from the opposite side with unchanged speed:

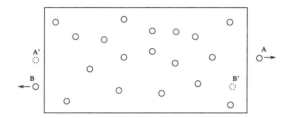

FIG. 3.5. *Periodic boundary conditions in a box.*

The total potential energy of the infinite system is given by the sum over all particles in the simulation box *and* all the periodic replicas of the box itself:

$$V = \sum_b \sum_{ij} \frac{q_i q_j}{r_{b,ij}} \tag{3.7}$$

q_i are the particle charges (or masses) and b runs over box itself ($b = 0$) plus all its replicas. Here $r_{b,ij} = |x_i - x_j + X_b|$, where X_b is the centermass coordinate of the b-th box relative to the simulation box. It can be shown that the recirculating boundary conditions introduce an error of the order of $1/L$. Thus this type of boundary conditions is only adequate for very short-ranged interactions.

3.4.2 *Minimum image*

Here the idea is that each particle is allowed to interact only with all particles falling into a box centered upon the particle itself:

This improves considerably over the recirculating boundaries for weakly coupled systems with screened interactions, typically with a Debye length l_D below the box size L. For unscreened interactions, such as gravitational systems, or one-component plasmas, this is again still inaccurate.

3.4.3 *Ewald summation*

The method of choice to deal with boundary conditions for long-range interactions is the so-called Ewald summation. The Ewald summation technique is rather involved, and therefore we shall only sketch the major ideas behind it. The main task is to replace the slowly convergent series (3.7) with two fast-convergent ones, one in real space (short-range) and the other in Fourier space (long-range). The main observation is that pointlike charges generate a $1/r$ potential which collects two undesirable features: it is slowly decaying at infinity

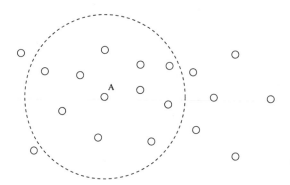

FIG. 3.6. *The particle A interacts only with particles inside the dashed box centered around itself.*

and singular at $r = 0$ at a time. The former property makes real-space series slowly convergent, while the latter causes problems to Fourier expansions in the high-wavenumber regions of reciprocal space. The idea is therefore to split the factor $1/r$ into a short and long range components, each of which hinerits only one of the two aforementioned problems. This splitting is performed by means of a compensating function $C(r)$, with the following properties:

$$C(r \to 0) \to 1$$
$$C(r \to \infty) \to e^{-r}$$

In other words, $C(r)$ is of Yukawa form, namely regular at short-distance and (at least) exponentially at large distances. With the aid of $C(r)$, we can define a pair of long and short-decay functions:

$$L(r) = \frac{1 - C(r)}{r}$$
$$S(r) = C(r)/r$$

such that $L(r)$ is long-ranged, but no longer singular around the origin, whereas $S(r)$ is still singular around the origin, but short-ranged. As a result, S can be treated by real-space methods (minimum image for instance) and L can be summed in Fourier space. A prototypical compensating function is the exponential $C(r) = e^{-r/l}$, where l is a typical screening length (for instance the Debye length in neutral plasmas). The physical interpretation is that a point-like charge $\delta(x - x_i)$ is dressed with a counter-acting cloud of charge $-W(x - x_i; l)$ such that $\delta - W$ represents a screened charged giving rise to a short-range potential. On the other hand, the cloud W itself generates a long-range potential with regular behaviour for $r < l$. As a result, the compensating function technique corresponds to the identity:

$$\delta = (\delta - W) + W$$

This general idea is then applied to the total potential energy expansion (3.7). The first step is to take a Fourier transform of (3.7)

$$V = \sum_K \sum_{ij} q_i q_j e^{iKr_{ij}} K^{-2}$$

where the wavenumber K runs over the periodic replicas. The divergence at $K = 0$ is only apparent since, for charge-neutral systems, the condition $\sum_i q_i = 0$ rules the case $K = 0$ out of the summation. If, like for the case of gravitation, the system is not charge-neutral, a compensating term of the form $\sum_{ij} q_i q_j \sum_{R \neq 0} 1/R$ must be subtracted. The point of Fourier-transforming the total energy is to turn slow convergence due to large r into a slow convergence due to large K. The latter, ultraviolet (large K) divergence (or slow convergence) is then tamed by replacing point-like particles with finite-size (dressed particles) of size h:

$$\delta(x - x_i) \rightarrow W_h(x - x_i)$$

where W are the shape functions centered around the particle position x_i, typically a Gaussian in the Ewald's method. The (not straightforward) result is the following expression:

$$V_{Ewald} = \sum_{K \neq 0} \sum_i |q_i e^{iKr_{ij}}|^2 \frac{1}{K^2} e^{-K^2/4h^2} + \sum_{i,j} q_i q_j \frac{1}{r_{ij}} Erfc(hr_{ij}) - h/\sqrt{\pi} \sum_i q_i^2$$

Now, the first sum converges rapidly due to the exponential, whereas the second term is short-ranged, and consequently it can be treated with the minimum image method. As a result, what one has to do is to replace the Coulombic potential $1/r_{ij}$ with the corresponding Ewald potential V_{Ewald} given above.

3.5   References

1. S. Pfalzner, P. Gibbon, Many-body tree methods in physics, Cambridge Univ. Press, 1996.
2. Leslie Greengard, The numerical solution of the N-body problem, Computers in Physics, March/April 1990, p.142.
3. J. Barnes, P. Hut, A hierarchical O(NlogN) Force-Calculation Algorithm, Nature, 324, 1986, p. 446.

4

GRID-BOUND PARTICLES: LATTICE GAS AND LATTICE BOLTZMANN

In this lecture we shall offer an account of Lattice Gas Cellular Automata and Lattice Boltzmann methods, which represent two remarkable examples of particle methods in which particle dynamics is forced to live in a discrete lattice. Despite this severe limitation, these methods prove capable of describing complex phenomena, such as those associated with the dynamics of real fluid flows.

4.1 Introduction

In the previous chapters we have seen techniques in which particles interact either directly (particle-particle,PP) or through the intermediate of a mesh-based field (particle-mesh,PM). In both cases particles were free to take any position (compatible with boundary conditions) in configuration space. We have also seen that placing the interactions on the lattice yields significant computational benefits. However, these benefits come at the expense of some accuracy, which is limited by the need of transferring information from/to the grid to/from the particle positions. Further efficiency can be gained by constraining particle motion to occur only at lattice positions (grid-bound models) simply because in this *particle-in-mesh (PIM)* models no interpolation is needed. The price to pay is that relevant physics may be altered by this drastic cut of degrees of freedom. This is why PIM particle methods are generally used as hyperstylized cartoons of a real physical system, with the main intent of capturing the universal aspects of the physics, without aiming at getting the hard numbers. However, in a few special cases, they do provide a quantitatively correct picture of the relevant physics. As an example in this class, we present here a remarkable instance of PIM technique, known as Lattice Gas Cellular Automata (LGCA), which is devised to solve the complex equations of fluid dynamics. A very successfull, (in fact, more successfull) spinoff of the LGCA technique, known as Lattice Boltzmann method is also briefly discussed.

4.2 Lattice gas

Let us begin by considering a regular lattice with hexagonal symmetry such that each lattice site is surrounded by six neighbors identified by a corresponding set of connecting vectors $\vec{d_i} \equiv d_{ia}$, $i = 1, 6$, the index $a = 1, 2$ running over the spatial dimensions. Each lattice site hosts up to six particles with the following prescriptions:

- *All particles have the same mass $m = 1$.*

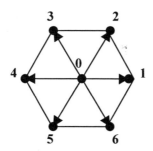

FIG. 4.1. The FHP hexagonal lattice.

- *Particles can move only along one of the six directions defined by the discrete displacements d_{ia}.*
- *In a time-cycle (made one for convenience) the particles hop to the nearest neighbor pointed by the corresponding discrete vector c_{ia}:*

$$c_{ia}dt = d_{ia} \qquad\qquad (4.1)$$

Both longer and shorter jumps are forbidden, which means all lattice particles have the same energy.

- *Two particles sitting on the same site cannot move along the same direction c_{ia} (Exclusion Principle).*

These prescriptions identify a very stylized gas analogue, which is known as Frisch-Hasslacher-Pomeau (FHP) lattice gas automaton, after his inventors in 1986.

In a real gas molecules move along any direction (*isotropy*), whereas here they are confined to a six-barred cage. Moreover, real molecules can move virtually at any (subluminal) speed, whereas here only six monochromatic beams are allowed. Amazingly enough, this apparently poor cartoon of true molecular dynamics has all it takes to simulate realistic hydrodynamics! With the prescription kit given above, the state of the system at each lattice site is unambiguously specified in terms of a plain *yes/no* option indicating whether or not a particle sits on the given site. That's all we need to know. This dichotomic situation is readily coded with a single binary-digit (bit) per site and direction so that the entire state of the lattice gas is specified by $6N$ bits, N being the number of lattice sites. It is expedient to introduce an *occupation number* n_i, such that

$$n_i(\vec{x}, t) = 0, \quad \text{particle absence at site } x \text{ and time } t \qquad (4.2)$$

$$n_i(\vec{x}, t) = 1, \quad \text{particle presence at site } x \text{ and time } t \qquad (4.3)$$

The collection of occupation numbers $n_i(\vec{x}, t)$ over the entire lattice defines a $6N$ dimensional time-dependent boolean field whose evolution takes place in a boolean phase-space consisting of 2^{6N} discrete states. This boolean field takes

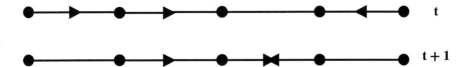

FIG. 4.2. Free-streaming in a discrete one-dimensional lattice.

the intriguing name of *Cellular Automaton* (CA), to emphasize the idea that not only space and time, but also the dependent variables (matter) take on discrete (boolean) values. The fine-grain microdynamics of this boolean field can*not* be expected to reproduce the true molecular dynamics to any reasonable degree of microscopic accuracy. However, as is known since Gibbs, many different microscopic systems can give rise to the same macroscopic dynamics, and it can therefore be hoped that the macroscopic dynamics of the lattice boolean field would replicate real-life hydrodynamic motion even if its microdynamics does not. In particular, as we shall see, its dynamics can be designed in such a way that the following quantities:

$$\rho = m < \sum_i n_i > \tag{4.4}$$

$$\rho u_a = m < \sum_i n_i c_{ia} > \tag{4.5}$$

where bracket denote ensemble averaging, evolve in space and time exactly as the mass density and current of a real fluid.

4.3   LGCA microdynamic evolution

Let us now prescribe the evolution rules of our CA. Since we aim at hydrodynamics, we should cater for two basic mechanisms:
- *Free-streaming*
- *Collisions*

Free streaming consists of simple particle transfers from site to site according to set of discrete speeds c_{ia}. Thus, a particle sitting at site $\vec{x}$ at time t with speed c_{ia} will move to site $\vec{x} + c_{ia}$ at time $t + 1$.

In equations:

$$n_i(\vec{x} + c_{ia}, t + 1) = n_i(\vec{x}, t) \tag{4.6}$$

This defines the discrete free-streaming operator S_i as:

$$S_i n_i \equiv n_i(\vec{x} + c_{ia}, t + 1) \tag{4.7}$$

The equation 4.6 reads then simply

$$(S_i - 1)n_i = 0$$

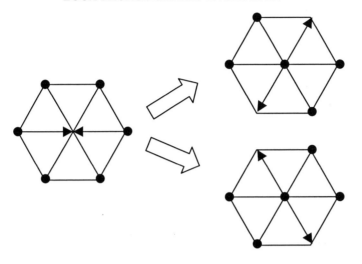

FIG. 4.3. A FHP collision with two equivalent outcomes.

This equation is a direct transcription of the Boltzmann free-streaming operator $D_t \equiv \partial_t + v_a\partial_a$ to a discrete lattice in which space-time are discretized according to the synchronous 'light-cone' rule:

$$\Delta x_{ia} = c_{ia}\Delta t \qquad (4.8)$$

The relation between the discrete and continuum streaming operators reads:

$$S_i = e^{D_i} \qquad (4.9)$$

where $D_i \equiv \partial_t + c_{ia}\partial_a$ is the generator of space-time translations along the i-th direction (sum over repeated indices is implied). Manifestly the speed magnitude c plays the role of the "light speed" in the discrete world, in that no signal can propagate faster than c in the lattice. Once on the same site, particles interact and reshuffle their momenta so as to exchange mass and momentum among the different directions allowed by the lattice.

This mimics the real-life collisions taking place in a real gas, with the crude restriction that all pre and post-collisional momenta are forced to live on the lattice. For all its simplicity, the FHP automaton proves able of reproducing most of the complexities of real fluid behaviour: the name of this magic is *symmetry and conservation*.

Consider the FHP collision depicted in Fig. 3. This collision shares two crucial features with a real molecular collision:

- It conserves particle number (2 before, 2 after)
- It conserves total momentum (0 before, 0 after)

These properties are a "conditio sine-qua-non" to achieve hydrodynamic behaviour once a sufficiently large group of particles is considered; this is what defines a fluid as opposed to a group of particles. It is a necessary but not sufficient condition, though. One may ask why not making things even simpler and consider for instance just a four-state automaton living in a square lattice. This is the so called Hardy-Pomeau-De Pazzis (HPP) cellular automaton, predating FHP of more than 10 years. The HPP automaton can also secure conservation laws. However, it fails to achieve a further basic symmetry of the Navier-Stokes equations, namely *rotational invariance*. Making abstraction of dissipative terms for simplicity, the hydrodynamic probe of isotropy is the tensor: $T_{ab} \equiv \rho u_a u_b + P \delta_{ab}$, where u_a is the fluid speed and P the fluid pressure. For a 2D Navier-Stokes fluid, this reads:

$$T_{xx} = P + \rho u^2, \tag{4.10}$$
$$T_{xy} = T_{yx} = \rho u v \tag{4.11}$$
$$T_{yy} = P + \rho v^2 \tag{4.12}$$

where u, v are the cartesian components of the flow field. This is an isotropic tensor because its components are invariant under arbitrary rotations of the reference frame. With HPP, we would obtain instead

$$T_{xx} = P + \rho(u^2 - v^2), \tag{4.13}$$
$$T_{xy} = T_{yx} = 0, \tag{4.14}$$
$$T_{yy} = P + \rho(v^2 - u^2) \tag{4.15}$$

Thus, the four HPP particles do not qualify for the status of *fluons*, i.e. elementary excitations whose collective dynamics reproduces macroscopic hydrodynamics. It is important to realize that this lack of isotropy cannot be cured by going to finer space-time resolutions, no matter how finer. In a mathematical language, this is because the continuous group of rotations $SO(1)$ is compact. In two-dimensions, the group of discrete rotations Z_6 (rotations by multiple of $2\pi/6$) can substitute for the continuous group $SO(1)$ while Z_4 cannot. More precisely, the lattice must generally provide enough symmetry to ensure the following tensorial identity (spatial indices are summed upon):

$$\left[\sum_i^z c_{ia} c_{ib} \left(c_{ic} c_{id} - \frac{c^2}{D} \delta_{cd} \right) \right] u_c u_d = u_a u_b \tag{4.16}$$

for any choice of the dyadic $u_a u_b$. Here D is the space dimensionality and z is the number of discrete speeds (the lattice coordination number).

This is a very stringent condition that weeds out most discrete lattices. Intuitively, the point is that as opposed to a scalar field, which corresponds to a point-like particle, fourth-order tensor fields correspond to extended objects, and consequently they sense more details of the space-time structure.

Now, back to the collision operator.

The effects of collisions on the occupation numbers is a change from n_i to n_i' on the same site

$$n_i' - n_i = C_i(\underline{n}) \qquad (4.17)$$

where $\underline{n} \equiv [n_1, n_2 \ldots n_b >$ denotes the set of occupation numbers at a given lattice site. To formalize the expression of C_i it proves expedient to label phase space via a bit-string $\underline{s} = [s_1, s_2 \ldots s_b >$ spanning the set of all possible (2^b) states at a given lattice site. For instance, numbering discrete speeds $1 - 6$ counter-clockwise starting from rightward propagation, $c_{1x} = 1$, $c_{1y} = 0$, the pre and post-collisional states read $\underline{s} = [100100 >$ and $\underline{s}' = [010010 >$ respectively. It is natural to define a *transition matrix* $A(\underline{s}, \underline{s}')$ flagging all permissible collisions from source state $\underline{s}$ to destination state $\underline{s}'$, $\underline{s} \to \underline{s}'$ as follows:

$$A(\underline{s}, \underline{s}') = 1 \text{ collision allowed} \qquad (4.18)$$

$$A(\underline{s}, \underline{s}') = 0 \text{ collision forbidden} \qquad (4.19)$$

"Allowed" means here compliant with conservation laws. The transition matrix obeys the *semi-detailed balance* condition:

$$\sum_{\underline{s}} A(\underline{s}, \underline{s}') = 1 \qquad (4.20)$$

meaning that every destination state necessarily comes from a source state within the phase-space of the automaton. This condition does not imply a one-to-one source-destination relationship, as it is the case of detailed-balance:

$$A(\underline{s}, \underline{s}') = A(\underline{s}', \underline{s}) \qquad (4.21)$$

The latter ensures micro-reversibility, while the former does not. Indeed, it is easily shown that a given pre-collisional FHP input state can land into more than one (actually, two) post-collisional output state compliant with conservation laws (head-on collisions can equally well rotate particle pairs $\pi/6$ left or right). Consequently, unlike HPP, the time evolution of the FHP automaton is no longer deterministic. In practice, the resulting lack of chiral invariance is easily disposed of by choosing either collisions with equal probabilities. Next, let us define the probability to have $\underline{s}$ as input state with occupation number $\underline{n}$:

$$P(\underline{s}, \underline{n}) = \prod_i^b n_i^{s_i} \bar{n}_i^{\bar{s}_i} \qquad (4.22)$$

Here overbar denotes complement to one, i.e. $\bar{n}_i \equiv 1 - n_i$ as it befits to fermionic degrees of freedom. Let us clarify with an example. The probability to occupy state $\underline{s} = [100100 >$ as an input string is given by $P[100100 >= n_1 \bar{n}_2 \bar{n}_3 n_4 \bar{n}_5 \bar{n}_6$.

This quantity is manifestly always zero, except when a particle with speed $\vec{c}_1$ AND a particle with speed $\vec{c}_4$ are sitting simultaneously on the node. In passing, we note that 'particle absence' is to all effects tantamount to 'hole

1

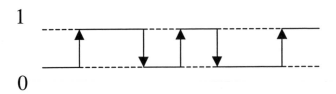

0

FIG. 4.4. The railway analogy of boolean computing. Only $0 \to 1$, $1 \to 0$ transitions can occurr, leaving the system on either upper or lower rails. Drifts are not allowed.

presence', echoing the particle-hole symmetry of fermionic matter . With these preparations, the collision operator can be formally recast in the traditional gain minus loss form

$$C_i = \sum_{\underline{s}\underline{s}'}(s_i' - s_i)P(\underline{s}, \underline{n})A(\underline{s}, \underline{s}') \tag{4.23}$$

The reader is encouraged to check that C_i is a "trit", namely a 3-state variable taking values -1 (annihilation), 0 (no action), $+1$ (generation).

The trit associated with the head-on collision depicted in Figure 4 is easily computed: $C_i = [-1, 1, 0, -1, +1, 0 >$. It is also readily checked that the sum over all discrete speeds yields identically zero, in fulfillment of the requirement of mass conservation. This is formally traced back to the identity:

$$n_i = \sum_{\underline{s}'\underline{s}} s_i P(\underline{s}, \underline{n})A(\underline{s}, \underline{s}') \tag{4.24}$$

which can be checked by direct computation.

The collision operator C_i obeys the remarkable property of preserving the boolean nature of the occupation numbers. This can be verified directly by noting that the Boolean-breaking occurrences, $n_i = 0, C_i = -1$ or $n_i = 1, C_i = 1$, can never take place. They are automatically ruled out by the fact that if $n_i = 0$, the collision operator cannot subtract particles to the input state, and conversely, if $n_i = 1$ it cannot add them to the pre-collisional state. This closely evokes the well-known properties of fermionic annihilation/generation operators of second-quantization.

To sum up, the final LGCA update rule reads as follows

$$(S_i - 1)n_i = C_i \tag{4.25}$$

or, equivalently:

$$n_i(\vec{x} + c_{ia}, t+1) = n_i'(\vec{x}, t) \tag{4.26}$$

where all quantities have been defined previously.

The equations (4.25, 4.26) represent the microdynamic equation for the boolean lattice gas, the analogue of Newton equations for real molecules. This equation

constitutes the starting point of a Lattice BBGKY hierarchy, ending up with
the Navier-Stokes equations. At each level, one formulates a lattice counterpart
of the various approximations pertaining to the four levels of the hierarchy. The
remarkable point is that, notwithstanding the drastic reduction of microscopic
degrees of freedom, the Lattice Navier-Stokes equation can be made basically
coincide with its continuum counterpart. In the LGCA jargon, the lattice is
"erased" from the macroscopic dynamics.

The guiding light of this clever procedure are the fundamental conservation
laws of classical mechanics. In boolean terms:

- *Mass and momentum conservation*

$$\sum_i c_i = 0 \qquad (4.27)$$

$$\sum_i c_{ia} c_i = 0 \qquad (4.28)$$

- *Angular momentum conservation (rotational invariance)*

$$\sum_i c_{ia} c_{ib} c_{ic} c_{id} = \frac{bc^4}{D(D+2)} (\delta_{ac}\delta_{bd} + \delta_{ad}\delta_{bc} - \frac{2}{D}\delta_{ab}\delta_{cd}) \qquad (4.29)$$

As it stands, the microdynamic equation (4.25) represents the dynamics of a
$6N$-body system of hopping spins ("spin fluid") with local mass and momentum
conservation, a very rich and interesting model of non-equilibrium statistical
mechanics. That the large-scale dynamics of this model does indeed reproduce
the fluid-dynamic equations remains still to be proved. We shall only sketch the
main ideas, directing the reader fond of full details to the excellent monographs
available in the literature.

4.4   From LGCA to Navier-Stokes

The standard *bottom-up* procedure taking from many-body particle dynamics
(Newton-Hamilton) all the way up to continuum fluid-like equations proceeds
through the three formal steps familiar from classical statistical mechanics:

1. *From Newton-Hamilton to Liouville*
   The main assumption here is ergodicity, namely the probability for the sys-
   tem to visit a given region of $6N$ dimensional phase-space $\Gamma_N = [\vec{x}_1, \vec{v}_1 \ldots \vec{x}_N, \vec{v}_N]$,
   is proportional to the volume of the region. This shifts the focus from New-
   tonian trajectories to $6N$-dimensional phase-space fluids described by the
   N-body distribution function $f_N = f(\vec{x}_1, \vec{v}_1, \ldots \vec{x}_N, \vec{v}_N, t)$.
2. *From Liouville to Boltzmann*
   Upon integration over many-body $6(N-1)$ degrees of freedom one ends up
   with the lowest-level (one-body) of the hierarchy, the one-body distribution
   function. $f_1 = f(\vec{x}_1, \vec{v}_1, t)$.

3. *From Boltzmann to Navier Stokes*
   Upon the assumption of small departures from local equilibrium, the lowest order kinetic moments of the one-body distribution, namely density, current and kinetic tensor, can be shown to evolve in time according to the Navier-Stokes equations of fluid dynamics.

Exactly the same steps are involved in the process of deriving the lattice Navier-Stokes equations from the lattice BBGKY, with the notable caveat that lattice discreteness needs to be handled with great care since continuum symmetries (Galilean invariance, roto-translations, parity) are always at risk of being broken by the lattice discreteness. The most dangerous diseases resulting from the lattice discreteness are:

- Low collisionality
  The number of lattice collisions allowed by the conservation laws is generally much smaller than in continuum space. The result is a very high molecular diffusivity ν implying very strong dissipative effects.

- Lack of Galilean invariance
  Owing to the lattice discreteness, local equilibria cannot be described by discrete Maxwellians, but only by polynomials expansions thereof. These polynomial expansions break Galilean invariance.

- Statistical noise
  Due to the crude boolean representation, in order to obtain smooth hydrodynamic signals, substantial space-time averaging is required. As a result, LGCA simulations require extremely large lattices.

- Lack of ergodicity (spurious invariants)
  Due to lattice discreteness, the LGCA dynamics conserves additional invariants, besides the physical ones (mass, momentum, energy). The existence of these invariants breaks ergodicity and undermines the BBGKY ensemble-averaging procedure described previously.

These diseases were (partially) recovered in many ingenious ways, as described in full detail in the specialized literature. Here we move on to the discussion of some practical implementation issues.

4.5   Practical implementation

The main computational assets of the LGCA approach to fluid dynamics are:

- *Exact computing (round-off freedom)*
- *Virtually unlimited parallelism*

The boolean nature of the LGCA update rule implies that the corresponding algorithm can be implemented in pure boolean logic, without ever needing floating-point computing. This is very remarkable, since it offers a chance to sidestep a number of headaches associated with the floating-point representation of real numbers, primarily round-off errors. Among others, the boolean representation eases out the infamous problem of numerical drifts plaguing long-time simulations both in fluid dynamics and fundamental studies in statistical mechanics. To see this in a more detail, let us consider the head-on collision of Figure 4.2

The collision is encoded by the following logic statement:
"If there is a particle in state 1 AND a hole in state 2 AND a hole in state 3 AND a particle in state 4 AND a hole in state 5 AND a hole in state 6, then the collision occurs. Otherwise nothing happens."
All we need to put this plain logical statement in practice are the elementary boolean operations AND (logical exclusion), OR (logical inclusion), NOT (negation) and XOR (inclusive OR). The effect of these operations is usually represented by the so called "truth-table" (1=TRUE and 0=FALSE) which we report below for simplicity:

A	B	A.and.B	A.or.B	A.xor.B	.not.A
0	0	0	0	0	1
0	1	0	1	1	
1	0	0	1	1	0
1	1	1	1	0	

Table 1: Truth table of the basic boolean operators.

The statement *"the collision occurs"* corresponds to setting up a collision mask flagging collisional configurations with $M = 1$ and with $M = 0$ all the others:

$$M = n_1.AND.(\bar{n}_2).AND.(\bar{n}_3).AND.(n_4).AND.(\bar{n}_5).AND.(\bar{n}_6) \qquad (4.30)$$

where bar means negation. Once the collision mask M is set up, the post-collisional state

$$n_i' = n_i + C_i \qquad (4.31)$$

is obtained by simply 'XORing' the pre-collisional state with the collisional mask.

$$n_i' = M.XOR.n_i \qquad (4.32)$$

This simple procedure is applied simultaneously in lock-step mode to all lattice sites in the typical "blind-fold" fashion so dear to vector computers. More importantly, each site is updated independently of all others, thus making the scheme ideal to parallel processing. Aptness to massively parallel processing is probably worth a few additional words of comment. Roughly speaking, parallel computing works on the time-honored roman principle "Divide et Impera" (Divide and Conquer). To solve a large problem, first break it into small species, then solve each piece independently, and finally glue all pieces to together to produce the global solution. By doing so, a collection of, say P, processors would ideally solve a given problem at a fraction $1/P$ of the cost on a single computer.

4.6   Lattice Boltzmann

All LGCA problems can be cured by a drastic change in perspective, which consists in moving to a lattice Boltzmann approach. The key idea is to replace Boolean occupation numbers with the corresponding one-body discrete distributions

$$f_i(x, t) = <n_i>$$

where $< \dots >$ stands for some form of ensemble averaging. Being an average quantity, f_i is by definition free from statistical noise. The price is the loss of boolean computing, because the pre-averaged distributions f_i are no longer boolean numbers.

A few significant advantages realized by LBE are:

1. Local equilibria can be pre-assigned as a function of the local fluid density and speed. This restores Galilean invariance, destroys spurious invariants and yields a dramatic boost of efficiency.

2. Relaxation rates to local equilibria can be free-tuned via a single relaxation frequency ω. This permits to achieve much lower viscosities than LGCA.

The result is a very simple, elegant and powerful discrete-kinetic equation, known as lattice BGK (from Bathnagar, Gross and Krook):

$$f_i(\vec{x} + \vec{c}_i, t + 1) - f_i(\vec{x}, t) = -\omega \left(f_i - f_i^e \right) \qquad (4.33)$$

where f_i^e are local equilibria and ω is a relaxation rate (inverse time scale) to this local equilibria, The local equilibria take the following general form:

$$f_i^e = \rho w_i \left\{ 1 + \beta c_{ia} u_a + \frac{\beta^2}{2} \left[c_{ia} c_{ib} - \beta^{-1} \delta_{ab} \right] u_a u_b \right\}$$

where w_i is a set of weights normalized to unity and β is the inverse temperature. The relaxation rate ω controls the non-equilibrium dynamics of the system and, in particular, it fixes the kinematic viscosity of the fluid according to:

$$\nu = (\omega^{-1} - 1/2)/\beta$$

The LBGK is a fairly efficient Navier-Stokes solver. Moreover it can easily incorporate additional mesoscopic physics within generalized local equilibria. For these reasons, it has undergone a vigorous growth in the last decade.

4.7   References

1. J.P. Rivet, J.P. Boon, Lattice Gas Cellular Automata, Cambridge Univ. Press, 2000,

2. B. Chopard, M. Droz, Cellular Automata modeling of physical systems, Cambridge Univ. Press, 1998,

3. S. Succi, The Lattice Boltzmann equation, Oxford Univ. Press, 2001.

4. H. Chen, Discrete Boltzmann systems and fluid flows, Computers in Physics, vol.7, n.6, p.632, 1993.

4.8   Exercises

1. Write a LGCA program to compute a 2D flow in a rectangular pipe. The flow is forced by a constant volumetric force. Fluid speed is zero on the upper and lower boundaries and recirculates periodically at inlet and outlet boundaries.

2. Same with Lattice BGK.

5

EQUILIBRIUM MONTE CARLO METHOD

The Monte Carlo method is a major technique to compute integrals and solve differential and integro-differential equations in high-dimensional spaces. The key is to sample these high-dimensional spaces so as to produce an optimal covering of the "important" regions. In this Chapter we shall illustrate the basic ideas of the equilibrium Monte Carlo method, namely applications involving the calculation of equilibrium averages of physical quantities defined by high-dimensional integrals.

5.1 Generalities

The (equilibrium) Monte Carlo method is a technique for computing averages expressed by integrals over a huge number of dimensions (easily 10^{23} in statistical mechanics). The typical example are phase-space averages in statistical mechanics, which are given by Avogadro-dimensional integrals of the form:

$$< A >= Z^{-1}(\beta) \int A(z)e^{-\beta H(z)} dz \qquad (5.1)$$

where

$$Z(\beta) = \int e^{-\beta H(z)} dz \qquad (5.2)$$

is the partition function. In the above, z is the phase-space coordinate, $A(z)$ is a generic phase-space function, H is the N-body Hamiltonian and $\beta = 1/K_B T$ is the inverse temperature of the system. It is easily seen that if the integrand is significant over a region of size $s \sim \sqrt{T}$, in a N-dimensional domain of linear size L, the relative volume of the significant region is only $(s/L)^N$. This fraction becomes incredibly small even at moderate values of N. For instance, in the case of $N = 100$ hard-spheres at the freezing temperature, this number is about 10^{-260}! It is therefore clear that the N-dimensional integral (5.2) requires smart strategies to place quadrature points selectively in the right regions of phase-space. Since Monte Carlo applications depend on methods to sample from a probability distribution, we begin by discussing the basic ideas behind sampling techniques.

5.2 Sampling from a distribution

We discuss three basic methods for sampling from a given distribution $p(x)$, namely:

1. *Exact Inversion*
2. *Composition methods*

3. Hit/Miss (Reject/Accept)

Since the basic pre-requisite of any sampling procedure is the availability of a good sequence of (pseudo)-random numbers drawn from a uniform distribution in, say $[0, 1]$, we shall discuss this item first.

5.2.1   Uniform distribution

Uniform random generators should fulfill three basic requirements:

- *Long recurrence period*
(the distance between two occurrences of the same sequence of numbers)
- *Low correlation*
Typical correlators such as $C_m = \sum_n r_n\, r_{n+m}$ should decay fast with m (a good idea to check this is to plot r_{n+m} versus r_n and make sure that "cloudy" pictures with no regular patterns are obtained).
- *Fast*
Since long sequences are often needed in applications, the generation of each member of the sequence should be a fast operation.

Uniformly distributed random numbers are typically obtained via *linear sequential recursions* of the form:

$$i_{n+1} = mod(Ai_n + B, C), \quad i_0 = S, \quad n = 0, 1, N$$

where A, B, C, S are (large) integers and S is the so called "seed", the starting point of the sequence. The above recursion generates a sequence of pseudo-random integers $0 < i_n < C$, such that division by C produces the desired sequence of reals in the range $[0, 1]$. The right choice of the parameters A, B, C guarantees a long recurrence period, ideally $T_r \sim 2^W$, where W is the bit length of the computer word (truly random numbers have $T_r \to \infty$). For instance, on a 32-bit word computer, a good choice is:

$$A = 7^7, \; B = 0, \; C = 2^{31} - 1$$

The actual implementation of this simple rule may hide some traps depending on the way different compilers handle the inevitable 'overflows' $(i_n > C)$ associated with the above relation. A simple architecture-independent version guarding against dangerous truncations is provided below:

```
c ==============================
          function randu(iseed)
c From Tao Pang, p.47
c ==============================
          i2e30 = 2**30
          ia=16807
          ic=i2e30-1      ! ic=2**31-1, but 2**31 is a overflow
          ic=ic+i2e30

          iq=ic/ia
```

```
ir=mod(ic,ia)

ih=iseed/iq
il=mod(iseed,iq)

it=ia*il-ir*ih
if(it.gt.0) then
   iseed=it
else
   iseed=it+ic
endif

ranpang=iseed/float(ic)

return
end
```

The generation of high-quality uniform random numbers is crucial to any Monte Carlo simulation and the effects of poor randomness should be watched out very carefully.

5.2.2 Sampling by Inversion

Assuming the availability of a (pseudo)-random sequence in the range $[0, 1]$, we are often presented with the task of generating a new sequence distributed according to a different probability distribution function (pdf), $p(x)$. Let $p(x)$ be a generic probability distribution defined in the interval $a < x < b$, normalized to unity $\int_a^b p(x)dx = 1$ and let $P(x) = \int_a^x p(y)dy$ be its primitive. By definition, $P(a) = 0$, $P(b) = 1$. Assuming the inverse $Q \equiv P^{-1}$ of the primitive is known, the exact inversion procedure reads simply:

$$r_x = Q(r_y) \tag{5.3}$$

where r_y is a random number drawn from a uniform distribution in $[0, 1]$.

In fact, let $u(y)$ be the uniform distribution, defined as:

$$u(y) = 1, \ if \ 0 < x < 1$$
$$u(y) = 0 \ elsewhere$$

By definition, the probability $p(x)$ fulfills the following relation:

$$u(y)dy = p(x)dx$$

where y is the image of x under p, so that the probability of finding x in p is the same as of finding y in u. Integrating both sides from $[a, y]$ and $[0, x]$ respectively, with initial conditions $P(a) = 0$, we obtain:

$$y = P(x)$$

which means

$$x = Q(y)$$

q.e.d.

The geometrical meaning of this equation is made clear by the picture 5.2.2:

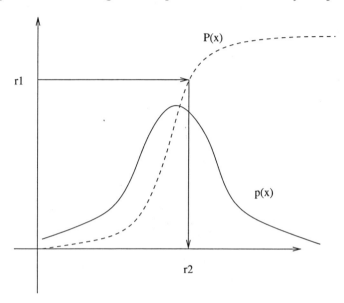

FIG. 5.1. *Sampling by inversion. From the uniform random number r_1 we obtain the random r_2 sampled from $p(x)$.*

A noticeable example of exactly invertible distributions is the exponential, $p(x) = e^{-kx}$:

$$P(x) = 1 - e^{-kx}/k =, \quad r_y = \frac{ln(1 - r_x)}{k}$$

Very often one needs to sample from Gaussian distributions, in which case the inversion procedure requires the inverse of the error function, not an easy function to work with.

The following trick (*Box-Mueller method*) saves the day.

Let $P(x, y) = e^{(x^2+y^2)/2} = G(x)G(y)$, where G denotes the normal Gaussian distribution (zero mean and unit variance):

Define: $x = \sqrt{2}Acos\phi$, $y = \sqrt{2}Asin\phi$, so that $dP = e^{-A^2}\frac{dA^2}{2}d\phi$. By letting $r_A = A^2$, we can sample A from the exponential distribution and $r_\phi = \phi$ from the uniform distribution in $[0, 2\pi]$. Finally, by trasforming back:

$$r_1 = \sqrt{-2\,ln(1 - r_A)}\,cos(2\pi r_\phi)$$
$$r_2 = \sqrt{-2\,ln(1 - r_A)}\,sin(2\pi r_\phi)$$

we obtain a pair of random numbers (r_1, r_2), both drawn from the (normal) Gaussian distribution.

Sampling by exact inversion yields perfect efficiency since each 'shot' delivers a valid random number. However, the need of knowing (either analytically or numerically) the primitive and its inverse, undermines the applicability of the method, especially in many-dimensions.

To this end, a more general method is needed. Before moving on to the discussion of this general method, we wish to emphasize that a large variety of distributions can be generated out of basic-block distributions, such as the exponential and Gaussian distributions, by using composition rules.

5.2.3 *Composition methods*

The probability distribution functions of two stochastic variables x and $y = f(x)$ are related as follows:

$$p_Y(y)dy = p_X(x)dx$$

namely,

$$p_Y(y) = p_X(g(y))/f'(g(y))$$

where $x = g(y)$ is the inverse of $y = f(x)$. These relations can be used to construct general pdf's starting from the basic ones (uniform, exponential, Gaussian). More generally, let us consider the joint distribution of a set of M stochastic variables z_j, $j = 1, M$, each of which is function of N independent stochastic variables x_i, $i = 1, N$:

$$z_j = f_j(x_1, \ldots x_N), \quad j = 1, M$$

A basic result of probability theory states that the joint distribution of the z's is given by

$$p(z_j) = \int dx_1 \ldots dx_N p(x_1, \ldots x_N) \prod_{j=1}^{M} \delta(z_j - f_j(x_1 \ldots x_N)) \qquad (5.4)$$

Useful instances of the above relation are:
Sum $z = x + y$:

$$p_Z(z = x + y) = \int p_X(x) p_Y(z - x) dx$$

Product $z = xy$:

$$p_Z(z = xy) = \int p_X(x) p_Y(z/x) \frac{1}{|x|} dx$$

where we assumed that x and y are independent (otherwise the integrals would not factorize). As an application, let us consider the sum z of two exponentially distributed variates. Application of the composition rule for the sum readily yields:

$$p_Z = ze^{-z}$$

Repeated application of this rule shows that the sum of N exponentially distributed variates is distributed according to:

$$p_Z = z^N e^{-z}/(N-1)!$$

namely the gamma distribution of order N. In practice, to sample from the gamma distribution of order N, one simply generates the sum:

$$z = -lnr_1 - lnr_2 \ldots - lnr_N$$

where r_i, $i = 1, N$ are drawn from the uniform distribution.

Consider now the ratio $z = y/x$ of two normally distributed variates. Application of the product rule to y and $1/x$, with some little algebra, shows that z is distributed according to a standard Lorentzian centered in $z = 0$:

$$p_Z = \frac{1}{\pi(1 + z^2)}$$

Both distribution can obviously be generalized to a Lorentzian centered in z_0 and width h, by a simple scale transformation:

$$z \leftarrow (z - z_0)/h$$

Composition methods significantly broaden the scope of inversion methods, in that they permit to generate a large variety of useful distributions. Even so, they do not easily extend to multidimensional distributions, for which a new class of methods is required.

5.2.4 Hit/miss Sampling

A fairly simple and general method to draw random numbers from virtually any distribution in one and more dimensions is the "hit/miss" technique, first proposed by von Neumann. The idea is simply to draw a pair of randoms and accept/reject them depending on the whether the corresponding point falls or not within the graph of the desired distribution:

1. Draw a pair (r_x, r_y),
   $a < r_x < b$, $p_{min} < r_y < p_{max}$
2. If $r_y < p(r_x)$: accept
   Else: reject

In other words, if the "trial" number r_y, falls below the value of the distribution $p(x)$ at $x = r_x$, it is accepted, otherwise it is rejected and a new toss is performed.

It is immediately appreciated that this procedure extends directly to the N-dimensional case by just drawing a $N + 1$-ple of randoms: $r_1, r_2 \ldots r_{N+1}$. The sour

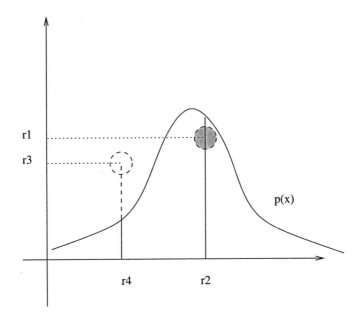

FIG. 5.2. *Hit Miss sampling. The random r_2 is accepted, while r_4 is rejected, because the point $(r4, r3)$ falls above $p(x)$.*

point is that some shots are necessarily rejected, so that perfect efficiency is lost. This is not a serious problem so long as one deals with 'well behaved', smooth distributions, evenly spread across their domain of definition. The method becomes unacceptably inefficient for "sneaky", highly-localized, distributions. Unfortunately, these are a commonplace in multi-dimensional applications.

To this purpose, "importance-sampling" techniques, as discussed in the sequel, are required.

5.3 Computing large-dimensional integrals

Suppose the task is to evaluate an integral of the form:

$$I = \int_{\Gamma_N} f(x)dx \tag{5.5}$$

where $x = [x_1, \ldots x_N]$ is a N-dimensional state vector in space Γ_N, and N a huge number (of the order of the Avogadro number). Equivalently, we seek the average of f over the N-dimensional space Γ_N:

$$<f> = \frac{1}{V} \int_{\Gamma_N} f(x)dx \tag{5.6}$$

where V is the volume of Γ_N. In principle, one may think of using standard quadrature formulae, such as Simpson, Gauss and others. This option is rapidly ruled out by the fact that if g is the number of quadrature points in one-dimension, the same accuracy in N dimensions requires g^N points! In addition, it is easy to see that such set of points is clustered around the surface of Γ_N, so that the interior domain is severly underrepresented. A better way to produce a uniform covering of Γ_N is to choose the integration points at *random*, i.e. generate a random sequence $[r_1 \dots r_N]$ and evaluate the average as the corresponding sum:

$$< f >~< f >_N = \frac{1}{N} \sum_{i=1}^{N} f(r_i) \tag{5.7}$$

Now we observe that the quality of the approximation $< f >_N$ depends on the degree of flatness of the function $f(x)$: with $f = constant$, a single quadrature point would produce the exact result, regardless of its location. For a generic function distributed moreless uniformly in Γ_N, the error decays like $1/\sqrt{N}$. In general, however, N-dimensional functions $f(x)$ are highly localized, and a strategic choice of the quadrature points is instrumental. The goal is of course to produce the shortest possible sequence of random numbers. The optimal strategy to achieve this goal is to place the quadrature points in region of phase-space where the integrand $f(x)$ contributes most to the integral in question. This strategy is generally known as *importance sampling*.

5.4 Importance sampling

To illustrate importance sampling, let us go back to a one-dimensional integral:

$$< f >= \frac{1}{h} \int_a^b f(x) dx = \int_a^b f(x) u(x) dx$$

where $h = b - a$ and

$$u(x) = 1/h \ \ a < x < b$$
$$u(x) = 0 \ \ elsewhere$$

is the uniform distribution. The above integral also rewrites as:

$$< f >= \int_{U(a)=0}^{U(b)=1} f[x(U)] dU$$

where we have set $dU = u dx$. It is apparent that the expression

$$f_N = \frac{1}{N} \sum_i^N f_i \equiv \sum_i w_i f_i$$

is nothing but a sampling from the uniform distribution, $w_i = dU_i = u_i dx = \frac{1}{h} \frac{h}{N} = 1/N$, being the (uniform) weight of each quadrature point. The operational meaning is just "throwing" points at random and summing up the contributions f_i so obtained (the typical image is the estimate of the size of a pond by counting the number of randomly thrown stones which fall within the pond). Suppose we happen to know a function $i(x)$ (importance function) which takes significant values only there where $f(x)$ is significant. With this knowledge, we can rewrite our average as:

$$< f >= \frac{1}{h} \int_a^b \frac{f(x)}{i(x)} i(x) dx$$

where it is clear that the rescaled function:

$$\hat{f}(x) \equiv \frac{f(x)}{i(x)}$$

is almost flat since f and i are very similar.

Therefore, we can write:

$$< f >= \frac{1}{h} \int_{W(a)}^{W(b)} \hat{f}[x(W)] dW$$

which means sampling the quadrature points no longer from a uniform distribution, but from a distribution W such that $dW/dx = i(x)$. This guarantees that shots are now 'guided' by $i(x)$ in such a way that most attempts fall within regions where $f(x)$ is substantial. The operational meaning of this expression is: throw a random number x_j in $[a, b]$ and accept it with probability i_j, namely, sample from $i(x)$.

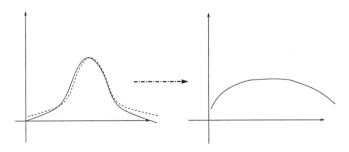

FIG. 5.3. *Importance sampling. Left: The distribution $p(x)$ (solid line) is approximated by an importance function (dashed line). Right: Note that the resulting rescaled function is much flatter than $p(x)$.*

A paramount instance of importance sampling is the Metropolis Monte Carlo technique discussed in the following.

5.5 Metropolis Monte Carlo

A basic task in equilibrium statistical mechanics is to take phase-space averages over the canonical ensemble:

$$< A >= \frac{\int e^{-E(x)/KT} A(x) dx}{\int e^{-E(x)/KT} dx} \tag{5.8}$$

The best method to compute this integral is provided by the Metropolis Monte Carlo algorithm, definitely one of the most important simulation tools ever. The task is to efficiently generate sequence of points distributed according to the canonical distribution:

$$p(x) = \frac{e^{-E(x)/KT}}{\int e^{-E(x)/KT} dx} \tag{5.9}$$

Since $p(x)$ is extremely localized in N dimensions, the trial points cannot be drawn from a generic random sequence.

The idea of the Metropolis MC algorithm is to sample from a Markov-chain in which the successor y of a given element x is not selected at random, but with a conditional probability $T(x, y)$ (transition kernel) which ensures that, after a while, the sequence of states does indeed belong to the canonical distribution (5.9). It can be shown that, in order to fulfill this property, the transition kernel must obey the semi-detailed balance condition:

$$p(x)T(x, y) = p(y)T(y, x) \tag{5.10}$$

This ensures that the probability of being in x and jumping therefrom to y is exactly the same as its inverse from y to x: a weak form of micro-reversibility. The micro-dynamic interpretation of this condition will be discussed in some more detail in the chapter devoted to Kinetic Monte Carlo. For now, we simply observe that semi-detailed balance implies the following relation:

$$\frac{T(x, y)}{T(y, x)} = \frac{p(y)}{p(x)} = e^{E(x) - E(y)} \equiv e^{-\delta E(x, y)} \tag{5.11}$$

where we have set $KT = 1$ for simplicity. This shows that the transitional probability depends only on the delta of the "landscape" function (energy) between the source and destination states x and y. The above condition still leaves a lot of freedom on the specific form of the kernel $T(x, y)$. Plausible choices are:

$$T(x, y) = \frac{1}{\tau} \frac{e^{-\delta E}}{1 + e^{-\delta E}} \tag{5.12}$$

or:

$$T(x, y) = e^{-\delta E}, \quad \delta E < 0 \tag{5.13}$$
$$T(x, y) = 1, \quad \delta E > 0 \tag{5.14}$$

In other words, downhill moves are always accepted, but eventually uphill moves are also accepted with probability $e^{-\delta E} < 1$. This tolerance to temporary energy-increasing "looser-moves" is essential to give the system the possibility of escaping from local minima. This is extremely important for many complex systems,

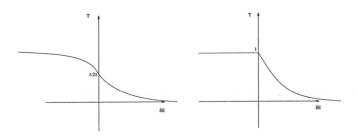

FIG. 5.4. $T(x, y)$ as a function of δE.

whose dynamics is frequently characterized by rough landascapes with many competing local minima.

The Metropolis Monte Carlo method has a very wide applicability to all sorts of optimization problems. In the following, we present a classical statistical mechanics application: the celebrated Ising model.

5.6 Application: the Ising model

One of the most popular applications of equilibrium Monte Carlo is the calculation of critical exponents of phase-transitions in condensed matter. A paradigm of phase transitions in magnetic systems is provided by the famous Ising model. Consider a system of spin variables distributed on a regular lattice, $s_i = [0, 1]$, $i = 1, N$, where $s = 0, 1$ mean spin-down and spin-up respectively. The total energy of this spin system is given by the following nearest-neighbor Hamiltonian:

$$H = -J \sum_{i,j=i\pm 1} s_i s_j - h \sum_i s_i \tag{5.15}$$

where J is the spin-spin coupling and h couples to an external magnetic field. If $J > 0$ (ferromagnetic) the interaction favours parallel alignement $(E = -J)$, which means cooperation with the external field to produce ordered configurations. In the opposite case $J < 0$ (anti-ferro), antiparallel alignement is favoured, the system opposes the effect of the external field, and tends to produce disordered configurations. On each pair of sites there are four possible configurations, as listed below:

```
spins     E


00,11    -J
01,10    +J
```

The main observables associated with the magnetic lattice are the mean energy and magnetization:

$$e = E/N = \sum_{ij} H_{ij}/N = -J \sum_i s_i (s_{i+1} + s_i)$$

$$m = M/N = \sum_i s_i/N$$

At high-temperature, the energy minimum $E = -2JN$ is attained by a zero-magnetization disordered phase, such as, for example, the one listed below for a linear chain of ten spins:

1001010110

Below the critical temperature T_c, the ordered phase becomes unstable (any minor perturbation h would drive the system away from it) and two broken-symmetry ordered phases $(m = \pm 1)$ become energetically favourable, simply because the system develops a virtually unbounded susceptibility to alignement/antialignement with the external perturbation.

1111111111 $\leftrightarrow$ 0000000000

This is the typical situation of second-order phase-transitions epitomized by the Landau-Ginzburg double-well potentials:

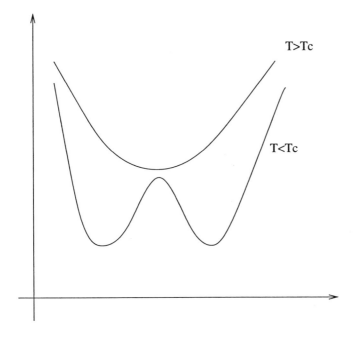

FIG. 5.5. *Double-well potential for $T < Tc$, single well above T_c.*

The task of the Monte Carlo simulation is to walk through configuration space until the system trajectory samples the ground state distribution. To this purpose, the Metropolis scheme proceeds as follows:

1. *Start from an initial spin configuration* [s]

2. *Move the spin configuration* $[s] \rightarrow [s']$
3. *Compute the corresponding energy change:*
   $$\delta E \equiv E[s'] - E[s] = -J \sum_i \delta s_i (s_{i+1} + s_{i-1})$$
4. *Accept the move with probability:* $p = min[1, e^{-\delta E / KT}]$.

If the moves satisfy semi-detailed balance, we are guaranteed that after some transient, the spin configurations belong to the desired canonical distribution. The question remains on *which* concrete spin moves (point 2 above) should be selected. Again, many choices are possible, but two remarkable examples are provided by Glauber "spin-flip" and Kawasaki "spin-flip-swap" moves.

5.6.1   *Glauber and Kawasaki moves*

The simplest and still popular form of spin move is the so-called Glauber spin-flip dynamics:

$$s_i \rightarrow s_i' = 1 - s_i$$

It is readily checked that Glauber dynamics yields the following energy changes (l, r denote the left and right neighbor spins respectively):

- Flip-down: $l1r \rightarrow l0r$: $\delta E = -4J$, l, r parallel
- Flip-down: $l1r \rightarrow l0r$: $\delta E = 0$, l, r antiparallel
- Flip-up: $l0r \rightarrow l1r$: $\delta E = +4J$, l, r parallel
- Flip-up: $l0r \rightarrow l1r$: $\delta E = 0$, l, r antiparallel

Glauber moves are very simple, but the corresponding dynamics is very slow because of the local nature of the move. In other words, Glauber dynamics corresponds to small steps in the Monte Carlo 'trajectory' across the multidimensional phase space. A faster dynamics, is realized by the Kawasaki 'flip-and-swap' moves:

$$s_i'(y) = 1 - s_i(x) \quad s_i'(x) = 1 - s_i(y)$$

where x and y are two distinct lattice sites.

FIG. 5.6. *Glauber (upper) and Kawasaki (lower) moves.*

Note that while Glauber moves do not leave the total magnetization of the system unchanged, the Kawasaki moves do. Thus, the two algorithms correspond to realizations of two different ensembles, "grand-canonical" (with independently assigned temperature and magnetic field) for Glauber, and "canonical" (with independent temperature and magnetization) for Kawasaki. While this would not matter with an infinite number of degrees of freedom, the rates at which the two ensembles reach equilibrium in finite-size systems can be pretty different, whence the importance of a judicious choice.

5.7 Advanced topics

Monte Carlo simulations have played a paramount role in the theory of phase-transitions. However, to tackle realistic problems, such as the computation of critical exponents free from finite-size effects, major technical advances are required, as compared to the simple background material presented here. These are described in full depth in the classical books given in the reference list. Here, we only mention two major problems, namely:

- *Critical slowing down*
  Near the critical temperature, the correlation time of the system grows virtually to infinity so that, in principle, one would need to wait an infinite amount of time in order to generate two independent configurations.
- *Finite-size effects*
  The critical exponents are generally highly sensitive to the size of the system (in principle, phase transitions are well defined only in the limit of infinite systems) again because the correlation length diverges in the proximity of a critical point.

5.7.1 *Cluster algorithms*

A powerful medicine against these problems are the so-called "global" Monte Carlo moves, in which a whole set (cluster) of properly selected spins are flipped simultaneously. The physical insight is that the divergence of the correlation length (or time) is due to the formation of macroscopic clusters of highly correlated spins. Critical slowing down can then be significantly counteracted by moves which prove able to efficiently destroy these clusters. The most representative of such global Monte Carlo schemes is the Swendsen-Wang cluster algorithm. The basic idea of the Swensen-Wang algorithm is to group spins according to the notion of "active bonds". No active bond can occur between two spins of opposite sign, whereas spins of the same sign are linked up with probability $p = 1 - e^{-2J/K_B T}$. Once clusters are in place, all spins in the cluster are flipped simultaneously, thereby realizing a "giant" step in the Monte Carlo trajectory. This permits to keep up with the divergence of the correlation length (time), thereby beating/mitigating critical slowing down.

5.8 References

1. D. Landau, K. Binder, Monte Carlo methods in statistical physics, Springer Verlag, 1997 (3rd edition).
2. K. Binder, D. Heermann, A guide to Monte Carlo simulations in Statistical Physics, Cambridge University Press, 2000.

5.9   Warm-up Ising program

```
c ====================================
            program ising2D
c ====================================
c from Binder p. 27
            include 'ising.par'
            parameter (D=2,mdim=2*2*D+1)
            dimension ip1(nx),im1(nx)
            dimension jp1(ny),jm1(ny)
            dimension ex(mdim)
c ------------------------------
            write(6,*) 'enter T'
            read (5,*) T
            write(6,*) 'how many sweeps?'
            read (5,*)  nsweeps

            Tc=0.22      ! critical temperature
c transition table
            beta = 1/T
            do 5 ide=1,mdim
             ex(ide)=exp(-beta*float(ide))
5           continue

            iseed=123457

            iout=0
            call init(s)
            call pbc(im1,ip1,jm1,jp1)

            beta = 1/T
            do it=1,nsweeps
             mag=0
             do j=1,ny
              jm=jm1(j)
              jp=jp1(j)

              do i=1,nx
               im=im1(i)
               ip=ip1(i)

               ide=2*s(i,j)*(s(ip,j)+s(i,jp)+s(im,j)+s(i,jm))
               re =ranpang(iseed)
               if(re.lt.ex(ide)) then
                s(i,j)=-s(i,j)
```

```
          endif
         mag=mag+s(i,j)
        end do
      end do
      write(6,*) it,mag
      write(7,*) it,mag
c output
      if(mod(it,nsweeps/10).eq.0) then
       iout=iout+1
       do 20 j=1,ny
       do 10 i=1,nx
        write(70+iout,*) i,j,s(i,j)
10        continue
20        continue
       endif

      end do
      stop
      end
c =================================
      subroutine init(s)
c =================================
      include 'ising.par'
c ---------------------------------
      iseed=7654321
      mag=0
      do j=1,ny
      do i=1,nx
       r=ranpang(iseed)
       if(r.gt.0.5) then
         s(i,j)= 1
       else
         s(i,j)=-1
       endif
       mag=mag+s(i,j)
       write(6,*) i,j,r,s(i,j)
       write(70,*) i,j,s(i,j)
      end do
      end do
      write(6,*) 'initial mag',mag
      pause

      return
      end
```

```
c  =======================================
             subroutine pbc(im1,ip1,jm1,jp1)
c  =======================================
             include 'ising.par'
             dimension ip1(nx),im1(nx)
             dimension jp1(ny),jm1(ny)
c  ---------------------------------------
             ip1(nx)=1
             do i=1,nx-1
              ip1(i)=i+1
             end do

             im1(1)=nx
             do i=2,nx
              im1(i)=i-1
             end do

             jp1(ny)=1
             do j=1,ny-1
              jp1(j)=j+1
             end do

             jm1(1)=ny
             do j=2,ny
              jm1(j)=j-1
             end do

             return
             end
```

5.10 Projects

1. Evaluate the integrals $I_n = \int_{-n}^{n} exp(-x^2)dx$ with $n \gg 1$ using importance sampling. Compare efficiency with a standard quadrature rule.
2. Using the 2D Ising program given above, write a computer code to compute the critical temperature of the 3D Ising model, using Glauber moves.
3. Same with Kawasaki moves.

6

QUANTUM MONTE-CARLO

The Monte Carlo method was generated to address complex problems in classical statistical mechanics and non-equilibrium transport theory. In the recent years, however, the method has gained high status also for the simulation of quantum systems, and particularly for the calculation of the ground-state and excited states of quantum many-body systems, a paramount issue in condensed-matter, material science and biology. In this chapter, we shall present a brief account of the main ideas behind the Quantum Monte Carlo (QMC) method.

6.1 Quantum many-body problems

The primary goal of QMC is the computation of the ground state, and lowest excited states, of the N-body Schroedinger equation. Mathematically, this amounts to solving the linear eigenvalue problem:

$$H\Phi(x) = E\Phi(x) \tag{6.1}$$

where

$$H = -\frac{\hbar^2}{2m}\sum_{i=1}^{N}\nabla_i^2 + \sum_{i,j>i=1}^{N} V(x_i, x_j)$$

is the N-body Hamiltonian in $3N$ dimensional coordinate space $x = (x_1, x_2, \ldots x_N)$. Like any ordinary eigenvalue problem, the idea is to find the lowest eigenvalue E_0 and the corresponding ground-state wavefunction Φ_0. Unlike ordinary eigenvalue problems, though, the monster dimensionality of the problem makes standard matrix algebra linear eigenvalues solvers virtually unviable. Therefore, qualitatively new methods are required.

We shall consider three popular instances of QMC:

- *Variational Monte Carlo (VMC)*
- *Diffusion Monte Carlo (DMC)*
- *Path-Integral Monte Carlo (PIMC)*

which we now discuss in some detail.

6.2 Variational Monte Carlo

The Variational Monte Carlo (VMC) method is a minimization technique built upon the variational principle for the Schroedinger equation. This principle states that the energy functional:

$$E[\phi, \phi] = (\phi, H\phi)$$

where $(,)$ denotes scalar product in Hilbert space L^{3N}, is minimized by the ground-state wavefunction $\Phi_0(x)$. As a result, a useful strategy to compute the ground state is to introduce a family of trial-functions $\phi(x, p)$, depending on the parameter p, compute the corresponding energy

$$E(p) = \int \phi^*(x, p) H \phi(x, p) dx / \int \phi^*(x, p) \phi(x, p) dx \qquad (6.2)$$

and finally look for the value $p = p^*$ such that E^* is minimum within the parameter space. The advantage is obvious: by choosing a low-dimensional parameter vector $p \equiv (p_1, p_2 \ldots p_M)$, $M << N$, the search of the minimum is conducted in a much smaller (non-linear) functional space of dimension $M << N$. The price is that the corresponding minimum E^* is only an approximation to the true minimum E_0. The idea is that physical intuition should help keeping the error

$$e(M) = |E^* - E_0|$$

below an acceptable threshold even though M is very small. Clearly, the success of VMC heavily leans upon a "good" choice for the trial wavefunction $\Phi_T(x, p)$.

The link with Monte Carlo procedures is as follows:

Rewrite the energy functional as:

$$E(p) = \int w(x, p) e(x, p) dx \qquad (6.3)$$

where (normalizing to unity):

$$w(x, p) = \Phi_T^*(x, p) \Phi_T(x, p)$$

is the local "particle" density, and

$$e(x, p) = \Phi_T^* H \Phi_T / \Phi_T^* \Phi_T$$

is the corresponding local energy. The above expression shows that $E(p)$ is given by a $3N$ dimensional integral, so that the core of the VMC is the evaluation of this high-dimensional integral. This is the typical task for a classical Monte Carlo quadrature, with importance function $w(x, p)$. The schematic VMC procedure is as follows.

For a given value of the parameter p:

1. *Initialization:*
   Choose a set on N_W walkers distributed at random in $3N$ dimensional coordinate space: $x \equiv [x_1, \ldots x_{N_W}]$.

2. *Monte Carlo move:*
   Move the generic k-th walker with a random displacement r_k: $x'_k = x_k + r_k$, so that a new state vector x' is generated.

3. *Metropolis test on the move:*

If $q \equiv \Phi_T[x']/\Phi_T[x] > 1$: accept the down-hill move,

Else: Accept the up-hill move, with probability q

This procedure "guides" the set of walkers (the quadrature points of the MC procedure) towards high-score regions identified by a substantial value of the importance function $w(x,p)$.

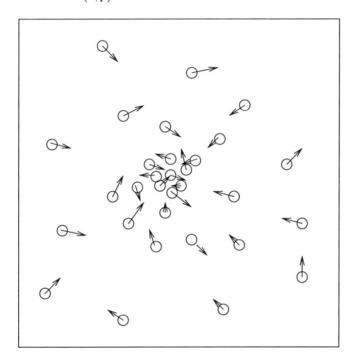

FIG. 6.1. *The random walkers tend to accumulate in those region where the integrand is most significant.*

Repeating the same procedure over a sequence of N_P parameter values $p_1, p_2 \ldots p_{N_P}$, delivers a corresponding sequence of tentative ground state energies $E_1, E_2 \ldots E_{N_P}$, whose minimum is then accepted as the current best approximation to the true ground state value.

Note however that since p is in general a $M-$dimensional vector, the search of this minimum becomes a non-trivial task for all but the smallest values of M.

6.2.1   Trial functions

As anticipated, good trial functions are crucial to the success of VMC. The optimal choice is a matter of good physical intuition and foreknowledge of the problem.

Useful trial functions are:

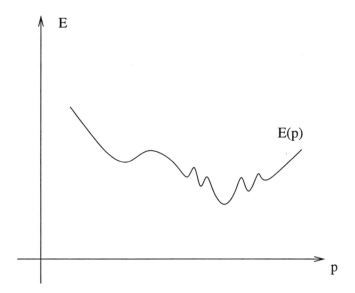

FIG. 6.2. *Energy of the ground state as a function of the parameter p.*

1. *1D Harmonic Oscillator:* $\Phi(x, p) = Ae^{-px^2}$,
   where A is a normalization constant.
   The local trial energy associated with $\Phi(x, p)$ is $e(x, p) = p + x^2(1/2 - 2p^2)$,
   which displays a minimum at $p = 1/2$, the exact solution.

2. *Helium atom :* $Ae^{-2(r_1 + r_2)}e^{\frac{r_{12}/2}{1 + pr_{12}}}$,
   where $\vec{r}_j$, $j = 1, 2$ are the positions of the two electrons and $r_{12} = |\vec{r}_1 - \vec{r}_2|$.
   Note that the parameter p only enters the correlation part of the trial
   wavefunction. Here, at variance with the harmonic oscillator, the best value
   of p does not provide the exact solution. This is generally signalled by a
   loss of sharpness of $E(p)$ around its minimum.

Multidimensional trial functions can be obtained by direct product of N one-dimensional ones. For most difficult (strongly interacting) cases, however, such direct products perform poorly and guessing the right trial function becomes a non-trivial task. This, besides high computational burden, is the Achille's heel of the VMC method.

6.3 Diffusion Monte Carlo

A more general method to evaluate the ground state of quantum many body systems is based on the well known equivalence between the Schroedinger equation in imaginary time and the diffusion and/or Fokker-Planck equation in real time. Upon performing a Wick rotation $t \rightarrow -it$, the Schroedinger equation turns into a diffusion-reaction equation with diffusivity $D = \hbar/2m$:

$$\partial_t \Phi = -H\Phi = D\Delta\Phi - V\Phi \tag{6.4}$$

The actual solution is a sum of decaying eigenfunctions:

$$\Phi(x,t) = \sum_n \phi_n(x)e^{-\omega_n t} \tag{6.5}$$

If the energy spectrum $E_n = \hbar\omega_n$ is well separated, after a sufficient long time, t_{relax}, the asymptotic solution decays self-similarly according to the smallest eigenvalue $E_0 = \hbar\omega_0$:

$$\Phi(x,t) \sim \phi_0(x)e^{-\omega_0 t} \tag{6.6}$$

The lowest eigenfrequency can then be computed as:

$$\omega_0 = \frac{ln[\Phi(x,t_2)/\Phi(x,t_1)]}{t_2 - t_1}$$

with $t_2 > t_1 >> t_{relax}$. In other words, the imaginary-time dynamics performs a natural projection upon the lowest-energy eigenstate. In actual practice a problem arises. Since the norm is not conserved, if the spectrum is not well separated, the signal may disappear before the smallest eigenvalue E_0 is well resolved. The remedy is to renormalize (almost) every time-step, so that the norm remains unitary all along the evolution. Formally this amounts to sustaining the wave-function with a compensating shift in the potential:

$$\partial_t \Phi = D\Delta\Phi - V\Phi + E_s\Phi \tag{6.7}$$

where E_s is tuned in such a way as to keep the norm constant. At the ground-state, we have $E_s = E_0$ since by definition, the shift energy must compensate exactly the ground-state decay. In practice, a typical DMC algorithm looks like follows:

1. Solve the diffusion equation
2. Compute the norm: $n(t) = (\phi^*, \phi)$
3. Rescale: $\phi \leftarrow \phi/\sqrt{n(t)}$.

The diffusion equation is solved by a classical MC sweep:

1. Initialization
   Choose a set on N_W walkers distributed at random in $3N$ dimensional coordinate space: $x \equiv [x_1, ...x_{N_W}]$.
2. Monte Carlo move
   Diffusion (Kinetic energy):
   Move a walker with a Brownian displacement
   $x'_k = x_k + d_k$,
   $d_k = \sqrt{2Ddt}\, r_k$, r_k, uniform random in $[0,1]$.
   Reaction: (Potential energy):

$V = 0.5 * (V(x) + V(x'))$

Compute $q = exp(dt(E_s - V))$

If $q < 1$: walker survives with probability q

Else: $int(q)$ new walkers are generated

This procedure is fine, except that if walkers come too close to each other, the potential energy contribution may lead to a demographic explosion.

6.3.1 Fokker-Planck approach

A useful remedy to population blow-up is to formulate the DMC procedure not on the wavefunction itself, but on the probability density defined with the aid of a (real) trial wavefunction $\phi_T(x)$:

$$\rho(x, t) = \phi_T(x)\phi(x, t)$$

Simple algebra shows that the imaginary-time Schroedinger equation turns into the following Fokker-Planck equation:

$$\partial_t \rho = \frac{1}{2} div[D\nabla\rho + U\rho] + (E_s - E_l)\rho \qquad (6.8)$$

where:

$$U = -2\nabla ln\phi_T$$

is the local drift, and

$$E_l = H\phi_T/\phi_T$$

is the local energy. The DMC procedure can then be applied to this Fokker-Planck equation, the advantage being that the branching step is now controlled by the factor $E_s - E_l$, which is no longer exposed to any divergence. In fact, at the ground-state this factor is just zero because at the ground state, $E_l = E_s = E_0$.

Again, we depend on the choice of a good trial function. However, this dependence is less critical than in VMC, since the dynamics of the walkers is no longer generic but proceeds according to the Schroedinger equation.

6.3.2 Excited states

Thus far we have illustrated computational strategies to evaluate the ground-state of quantum many-body systems. In many applications, excited states play also a strategic role (e.g, chemical reactions).

In principle, excited states can be obtained as follows.

Expand the initial wavefunction upon its eigenfunctions:

$$\phi(x, t) = c_0\phi_0(x)e^{-\omega_0 t} + c_1\phi_1(x)e^{-\omega_1 t} + c_2\phi_2(x)e^{-\omega_2 t} + \dots$$

Once $\phi_0(x)$ is known, we can compute the coefficient $c_0 = (\phi(x, 0), \phi_0)/(\phi_0, \phi_0)$, and form the initial shifted wavefunction:

$$\phi'(x, 0) \equiv \phi(x, 0) - c_0\phi_0(x)$$

Since the shifted wavefunction at time $t = 0$ does not project upon ϕ_0, the frequency ω_0 is never excited in the course of the evolution (this is a linear problem).

Consequently, the next excited eigenstate $\phi_1(x)$ is simply the ground-state of the shifted wavefunction ϕ'. As a result, the same DMC procedure can be applied to ϕ', to deliver the first excited eigenstate and eigenfunction. The practical problem with this procedure is its sequential character: first, each excited state requires the knowledge of all previous eigenstates, second, the entire procedure has to be repeated from scratch each time. Last but not least, ϕ' is no longer positive definite, which results in a significant complication for the sampling procedures (see section on Fermionic systems).

6.4 Path Integral Monte Carlo

A convenient approach to compute time-dependent (excited states) properties of quantum many body systems at finite-temperature is provided by the Path Integral Monte Carlo (PIMC) method.

The goal of the PIMC method is to compute the matrix elements of the evolution operator (in imaginary time):

$$P(t) \equiv e^{-Ht}$$

which defines the quantum N-body propagator (Green function):

$$G(x, y; t) = < x|e^{-Ht}|y > \tag{6.9}$$

where $|x>$ and $|y>$ stand for wavefunctions in 3N-dimensional real space, peaked at $q = x$ and $q = y$ respectively. Operationally:

$$G(x, y; t) = \int \delta(x - q)G(q, q'; t)\delta(q' - y)dqdq' \tag{6.10}$$

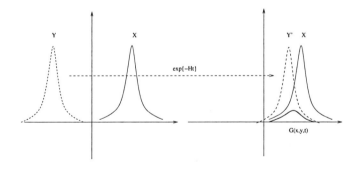

FIG. 6.3. *Graphical representation of the Green function.*

As is well known, the propagator of a d-dimensional quantum system is equivalent to the partition function of a $d + 1$ dimensional classical system at a temperature

$$T = 1/t$$

Since PIMC samples configurations with the appropriate Boltzmann weight, it permits to compute the expectation values of quantum systems at a finite temperature. Ground states are obtained in the limit of very long "evolutions", i.e. $t \to \infty$. In fact, based on the explicit eigen-expansion of the propagator:

$$G(x, y; t) = \sum_{jk} \phi_j(x) e^{-(\omega_j - \omega_k)t} \phi_k(y)$$

we have, for $t \to \infty$,

$$G(x, x; t) \simeq \phi_0(x)\phi_0(x)$$

namely the ground-state probability distribution.

6.4.1 *Time-slicing*

The evaluation of the propagator $P(t)$ at finite time is equivalent to knowing a finite time-span of the solution, which is not possible. However, for short enough time-spans, the path integral can be written as the product of quasi-exact short-time propagators. Consider a sequence of $M + 1$ instants, equispaced in time by the amount $h = t/M$ (time-slicing): $t_0 = 0, t_1 = dt, \ldots t_M = t$ and corresponding $M + 1$ coordinates x_m, with boundary conditions $x_0 = x$ and $x_M = y$.

Owing to the operator relation (strictly valid as $M \to \infty$):

$$e^{-tH} = (e^{-tH/M})^M$$

the finite-time propagator can be represented by a direct ordered product of short-time propagators:

$$P(x, y; t) = P(x, x_1)P(x_1, x_2) \ldots P(x_2, y)$$

If each sub-interval $dt = t/M$ is short enough, the corresponding propagator can be given an explicit expression, which is nothing but the Green-function of the Schroedinger equation:

$$G(x_m, x_{m+1}; h) = A e^{-h[V(x_{m+1/2}) + (x_{m+1} - x_m)^2/2h^2]}, \quad m = 0, M$$

where $h \equiv dt$, $x_{m+1/2} = (x_m + x_{m+1})/2$ and $A = 2\pi h^{3N/2}$ is a normalization constant. For the case $x = y$ we obtain:

$$P(x, x; t) \sim A^M \int Dx_0 \ldots Dx_M e^{-h \sum_m [V(x_{m+1/2}) + (x_{m+1} - x_m)^2/2h^2]}$$

Based on the Trotter-Suzuki splitting $e^{K+V} = e^K e^V + O([K, V])$, where K and V are the kinetic and potential energy operators, this formula is exact up to errors of order $O(h^2)$ and provides an operational recipe to evaluate the M-point path integral. Before we proceed to illustrate the numerics, let us notice that the above equation corresponds to the Lagrangian density of a system of $M + 1$ particles in $3N$ dimensions (or, equivalently $NM + N$ particles in 3 dimensions) interacting each other via a harmonic potential as well as with an external field $V(x)$. In other words the 3N-dimensional wavefunction $|x>$ is replaced by a system of $M+1$ particles in 3N-dimensions, or $NM+N$ particles in three dimensions. The corresponding path is conveniently thought as a series of N directed polymers in 3D, each constituted by $M+1$ monomers. For ground-state calculations $(x = y)$ the polymers close up into closed strings whose mean radius grows with the temperature of the system.

The path integral is then evaluated using Monte Carlo moves on this set of polymers.

6.4.2 Numerical evaluation of path integral

In order to actually compute the path integral, we set up M time-slices, each hosting N particles (walkers).

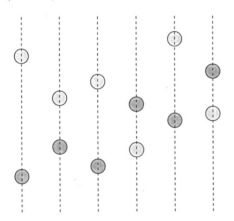

FIG. 6.4. *Path integration with two walkers and six time slices.*

A typical PIMC procedure would proceed as follows:

1. Initialization
   Put $NM + N$ particles in random positions x_n^m, $n = 1, N$, $m = 0, M$.
   Do while not finished:
   Select a time slice m at random
   Select a random particle n within slice m, x_n^m
   Move the particle to a new position with a random displacement: $x'^m_n = x_n^m + r$ and compute the new energy E'.
2. Metropolis test:
   Compute $p = e^{-h(E-E')}$
   Accept the move with probability $min[1, p]$
   End Do

What type of information can we extract from this procedure?

With a finite $t = Mh$, finite-temperature properties are obtained by averaging over the ensemble of Monte Carlo realizations. Ground-state properties are then obtained by making t very large (small temperature limit). The accuracy of the method depends on the size of the "time-step" h (at any given t) and the number of realizations. The former can be improved by more sophisticated representations of the short-time propagators. The latter by more efficient sampling. Full details can be found in the references provided herewith, as well as in the specialized literature on the subject.

6.5   Fermionic systems

Thus far we have been illustrating procedures in which the system wavefunction was implicitly assumed to be positive definite. This property is indeed satisfied by the ground state of bosons, but the same does not apply to fermionic systems. Indeed, as is well known, owing to Pauli's exclusion principle, many-body fermionic wavefunctions change sign whenever two particle coordinates take the same value (zero-hypersurfaces). To handle this problem, a number of very useful methods have been devised, none of which, however, attains a complete solution. Among the major ones, we mention the Fixed Node approximation and the Transient Estimations methods.

6.5.0.1   *Fixed-Node approximation*   The idea of the Fixed-Node approximation is simply to specify in advance the location of the hypersurfaces Σ_{ij} where particle i and j come to the same spatial location. Mathematically, this is defined by the condition

$$\Phi_0(x_1 \ldots x_i = x_j \ldots x_j = x_i \ldots x_N) = 0$$

where index 0 denotes ground-state. Of course, these hypersurfaces are not known in advance, but reasonable guesses can be made. The QMC simulations would then proceed as usual, with the only proviso of rejecting moves which take the walkers across a zero-hypersurface. Of course, the success of the fixed node approximation rests on a good choice of the zero-hypersurfaces.

6.5.0.2   *Transient Estimator method*   The idea of the Transient Estimator method is to decompose the fermionic wavefunction at imaginary time $t = 0$ into a symmetric and asymmetric components:

$$\phi_\pm = \frac{|\phi| \pm \phi}{2}$$

By definition, $\phi_+(\phi_-)$ is positive wherever ϕ is positive(negative), and zero elsewhere. So, they both represent genuine bosonic wavefunctions. The true wavefunction is simply the difference of the two:

$$\phi = \phi_+ - \phi_-$$

It can be shown that the imaginary-time evolution preserves the positivity of both $\phi_\pm$, so that each of them can be tracked with the methods discussed thus far.

The main problem, however, is that since both distributions are bosonic, they obviously tend to a bosonic ground state in the long time limit. The fermionic ground-state is an excited state of the many-body Hamiltonian, and consequently, in the long-time limit, the bosonic ground-state dominates by a factor $e^{(E_F - E_B)t}$, where E_B, E_F are the bosonic and fermionic ground-state energies respectively. The result is that the fermionic groundstate wavefunction is given by the difference of two large and nearly equal bosonic wavefunctions. This situation is exposed to very severe numerical inaccuracies due to cancellation errors, and consequently one must hope to detect a sizeable (transient) fermionic signal before this gets totally overshadowed by the bosonic ones. Ways to mitigate these cancellation errors represent a very active area of modern condensed matter research.

6.6   References

1. J. M. Thijssen, Computational Physics, Cambridge Univ. Press, 1998
2. D. Ceperley, B. Alder, Quantum Monte Carlo, Science, February 7, page 555, 1986.

6.7   Projects

Compute the ground state of the one-dimensional harmonic oscillator using VMC, DMC and PIMC. Compare the computational efficiency of the three methods.

7

KINETIC MONTE CARLO

In this lecture we describe real-time Monte Carlo methods (kinetic Monte Carlo), a technique especially suited to describe the dynamics of complex systems at a mesoscopic scale, i.e. an intermediate scale between atomistic and continuum representation.

7.1 Generalities

Kinetic Monte Carlo (KMC) is the tool of choice for time-dependent, high-dimensional transport phenomena involving multiple concurrent processes, taking place in complex geometries. Examples in point are:

- Particle transport (radiation, neutrons, electrons)
- High-energy experiments
- Nuclear reactor design
- Medical imaging
- Diffusion in solids
- Neutral plasmas
- ...

and many others. KMC takes the stage whenever geometrical complexity (real-life shaped objects) and/or physical complexity (multiple concurrent processes, possibly occurring at disparate space-time scales), impair the use of partial differential equations.

7.2 The Master equation

The system is described in probabilistic terms, via $P(x,t)$, the probability to find the system at state x at time t. Here x collects all the attributes of the system and is generally a very high dimensional state variable for instance, particle position, energy, angular momentum, size and so on. The time evolution of $P(x,t)$ is given by a balance equation of the generic Gain minus Loss form:

$$D_t P = G(P) - L(P) \tag{7.1}$$

where

$$D_t = \partial_t + \dot{x}\partial_x$$

is the total derivative along the system trajectory, and Gain and Loss are given in terms of transition kernels:

$$D_t P = \sum_y T(x,y)P(y) - \sum_y T(y,x)P(x) \tag{7.2}$$

The transition kernels collect a whole series of K physical processes, labelled here by index k,

$$T(x,y) = \sum_{k=1}^{K} T_k(x,y)$$

The transition kernels $T_k(x,y)$ are typically quite complicated expressions which are often given in tabular (non-analytic) form (e.g. nuclear databases). Each process can be associated with a specific 'event' in the particle history, which is why KMC can be categorized in the family of Event-Driven simulation methods.

As an example: A neutron entering a slab reactor is slowed down by collisions with atoms of the host medium until it is eventually absorbed by U-235 fissile atoms. The corresponding excited complex can either relax by emissio of gamma rays, or produce a fission event.

```
              scattering
neutron --->                        fission
              absorption --->
                              gamma emission
```

This is conveniently represented by a history-tree of events.

7.3   General properties of the transition kernels

Regardless of their specific form, which can be quite complicated, the transition kernels must obey a few general rules. These make the object of a growing branch of applied mathematics and statistical mechanics known as Transition State Theory.

7.3.1   Detailed balance

If each transition $x \to y$ has an equi-probable reverse transition $y \to x$, the system is micro-reversible:

$$T(x,y) = T(y,x)$$

This is also known as a detailed balance property. In many systems this condition is not satisfied. For instance, for activated processes, the probability to jump from x to y depends on the energy barrier separating the two states.
The result is (in units of $k_B T$):

$$\frac{T(x,y)}{T(y,x)} = e^{\Delta E(x,y)}$$

where $\Delta E(x,y)$ is the energy barrier separating state x from state y. In general, $\Delta E(x,y) \neq \Delta E(y,x)$, unless $E(x) = E(y)$.

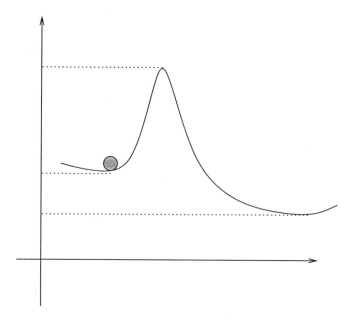

FIG. 7.1. *Activation energy barrier. To move to the bottom of the right valley, the system must overcome the top of the mountain.*

7.3.2 *Semi-detailed balance*

By summing over all possible destination states we obtain the outcome rate (losses per unit time) from state x:

$$\sum_y T(x,y) = O(x)$$

By summing over all possible source states we obtain the income rate (gain per unit time) of state y:

$$\sum_y T(y,x) = I(x)$$

Semi-detailed balance states the equality:

$$O(x) = I(x)$$

Note that detailed balance implies semi-detailed balance but not viceversa, i.e., semi-detailed is a weaker property. The outcome and income rates are clearly related to gain and loss terms:

$$G(x) = I(x)\bar{P}(x), \ L(x) = O(x)\bar{P}(x)$$

where $\bar{P}(x) = \sum_y p(x)T(x,y)/\sum_y T(x,y)$, is the mean probability to realize a state y jumping into x.

7.3.3   Dynamic equilibrium

The condition for (dynamic) equilibrium is

$$G(x) = L(x)$$

which is,

$$\frac{P^{eq}(y)}{P^{eq}(x)} = \frac{T(x,y)}{T(y,x)}$$

For separable kernels, $T(x,y) = U(x)V(y)$, this yields

$$P^{eq}(x) = V(x)/U(x)$$

Relevant cases are systems with detailed balance: $V = const.\ U$, $P^{eq}(x) = const.$, and activated systems: $P^{eq}(x) = e^{-E(x)/K_b T}$.

7.3.4   Link to Boltzmann and Fokker-Planck equation

Under the appropriate limits, the master equation generates the most important kinetic equations, such as the Boltzmann and Fokker-Planck equations. The link to the Boltzmann equation is very direct, since this equation is already in the form "Streaming=Gain-Losses". The Fokker-Planck equations is obtained by expanding the transition kernels in powers of the interstate separation:

$$r = |y - x|$$

The Fokker-Planck equation results from second order expansions in r, i.e. in the limit of "small jumps" (Brownian motion, as opposed to hard-core scattering). Long jumps can in principle be accounted for by higher-order expansions (Kramers-Moyal). However, these are much less useful, because the corresponding equations do not comply with the second principle of thermodynamics. Numerical solution of the Master equation via Kinetic Monte Carlo is obviously free of any such restriction.

7.4   Solving the integro-differential Boltzmann equation

Having discussed the general properties of the transition kernels, we move on to the illustration of how to solve the Master equation in practice. Leaving aside the total derivative, which is handled by standard particle pushing methods, we focus on sampling techniques to evaluate the integral kernels. For the sake of concreteness, we shall refer to the following transport problem. A beam of energetic neutrons is emitted by a localized source within a slab of material enriched with fissile Uranium 235. The neutrons are slowed down by inelastic scattering with the atoms of the host medium, and once they are thermalized at room temperature, they are absorbed by Uranium atoms at a much higher probability. As a result of the absorbtion, Uranium undergoes a fission event with probability p_F and releases $\nu > 1$ high-energy neutrons with speed v_F (peaked at $2MeV$ with tails up to $10MeV$). This chain of the event is described by the following transport Boltzmann equation in integro-differential form.

$$D_t f = -(\omega_a + \omega_s)f + \int [\Omega_s(v', v) + \nu\Omega_f(v, v')]f(v')dv' + S(v)\delta(x)\delta(t)$$

Here $f \equiv f(x, v, t)$, ω_k, $k = s, a$ is the characteristic frequency for absorbtion/scattering events, $\Omega_{s/f}(v, v')$ is the scattering/fission kernel from v to v' and Finally, S is a localized, impulsive source.

7.5   Sampling techniques

The probability of event k in a (small) time lapse dt is:

$$p_k = n_T v \sigma_k(v) dt \tag{7.3}$$

where n_T is the density of target atoms, v the neutron relative speed to target atoms, and σ_k the differential cross section of the k-th process (length squared). This defines a corresponding time and lengthscale for the process k,

$$\tau_k = 1/n_T v \sigma_k, \ \lambda_k = v\tau_k$$

These are the basic quantities to be used in the sampling procedure which evaluates the right-hand side of the Boltzmann equation. Very often in transport theory, the most used quantities are the global cross sections

$$\Sigma_k = n_T \sigma_k$$

which have the dimension of an inverse length. With this definition, the probability of event k in a time interval dt is simply:

$$p_k = v\Sigma_k(v)dt \equiv dl/\lambda_k = dt/\tau_k \tag{7.4}$$

where $dl = vdt$ is the path length travelled by the particle in a time lapse dt.
  Two major sampling techniques are:

- Differential sampling
- Integral sampling

7.5.1   *Differential sampling*

With K total processes we can form the ordered sequence of cumulative probabilities $P_m = \sum_{k=1}^{m} p_k$, $P_K = 1$. Once this is formed, we draw a random number r uniformly distributed between $[0, 1]$, and locate it within the sequence $\{P_m\}$. If r falls within, say, $[P_m, P_{m+1}]$, the m-th event is selected.

1   2   3   K

```
|----|-------|--r----|  ...... |

0   p1      p1+p2              1
```

Sketch 6.1: If Pm<r<Pm+1: event m occurs

This is called differential sampling, because it is based on the differential cross sections σ_k. In practice this procedure is convenient only if the number of processes is small, for otherwise the location of the m-th event becomes tedious and expensive. A better alternative is integral sampling, discussed below.

7.5.2   Integral sampling

Consider a particle at position $x = 0$; the probability for this particle to undergo event k upon flying over a distance l is given by:

$$P_k(l) = 1 - e^{-l/\lambda_k}$$

and conversely,

$$Q_k(l) = e^{-l/\lambda_k}$$

is the probability to fly a distance l without undergoing the k-th event (survival). These expressions only hold in the limit where λ_k can be assumed uniform in space (otherwise the integral $\int_0^l dx/\lambda(x)$ is required). With K competing processes, the above expressions generalize to $P = 1 - \prod_k Q_k$. As a result, all we have to do is sample from a set of exponential distributions. In practice, we form the sequence:

$$l_k = -\lambda_k \, lnr_k$$

and then move the particle over the minimum distance $l_m = Min_k[l_k]$, where event m is performed. Although equivalent to differential sampling, integral sampling is more straightforward and convenient from the computational point of view.

More practical yet, is to move the particle up to a distance

$$l = -\lambda_T \, lnr$$

and then use differential sampling to decide which event, say absorption or scattering, takes place at the endpoint of the flight. If it is absorption, the particle history ends up and a new particle history is initiated. If it is scattering, we must select a new particle velocity according on the scattering kernel $\Omega(v, v')$. For instance, for the simple case of isotropic scattering

$$\Omega(v, v') = \frac{1}{4\pi v^2}\delta(v - v')$$

we select two random numbers for the scattering and azimuthal angles:

$$\theta = 2\pi r_1, \quad \phi = 2\pi r_2$$

and then compute $v'_x = v\sin\theta\cos\phi$, $v'_y = v\sin\theta\sin\phi$, $v'_z = v\cos\theta$. These define the new particle velocity in a local frame, with z axis aligned with the pre-scattering speed v. With this updated speed a new flight can be initiated.

7.6   Variance reduction

The general problem of any Monte Carlo procedure is to guarantee a reasonable statistical accuracy. This is not easy since statistical errors decay very slowly, essentially with the square root of the number of samples. As a result, variance reduction techniques are essential to any efficient use of the KMC method. The general idea is to classify particles according to their *importance*, in that one tries to keep in the simulation only those particles which make a significant contribution to the process under inspection. This is generally known as importance sampling.

7.6.1   Biased-source sampling

Consider a beam of gamma rays irradiating the left side of a slab and assume we want to compute the transmittance of the slab, namely the fraction of incident gammas which make it to the right side of the slab. Within the slab, gamma rays can undergo either anelastic scattering or be absorbed by host atoms with the consequent emission of an electron (photo-electric effect). Let $n(E)$ be the energy spectrum of the incident beam; since only most energetic gammas will make to the right side of the slab, it makes sense to distort the original spectrum into an artificial one,$n'(E)$, richer in high-energy gammas. By giving a correspondingly reduced weight

$$w'(E) = n(E)/n'(E)$$

such that $\int n(E)de = \int n'(E)w'(E)dE$, low-energy particles, which make a minor contribution to the ultimate score, are suppressed, whereas valuable high-energy gammas are enhanced. This simple trick can improve significantly the statistical confidence of the computed transmittance.

7.6.2   Biased-score sampling

On a similar vein, assume we want to compute the spatial distribution of the energy deposited by the scattered/absorbed gamma rays within the shield. Since absorption is an undesirable event for statistical accuracy, we may decide to suppress it altogether; i.e. particles can never be absorbed. However, at each given 'pseudo/absorption' they deposit a reduced score:

$$E(x) = E(x) + \epsilon\Sigma/\Sigma_a$$

where E is the energy score at position x within the slab and ϵ is the amount of energy released in an absorption event.

7.6.3   Splitting and Russian roulette

Splitting, originally due to von Neumann, is probably the most natural form of variance reduction. The idea is that 'clever' particles which make it to a given distance x receive a reward in the form of a pool of $b > 1$ newborn children, each with a fraction $1/b$ of the original weight of the mother particle. As a result, by laying down a certain number of splitting surfaces within the geometrical domain, one can maintain good accuracy down into the deeper regions. Of course, the opposite problem must be monitored too: the birth-rate must be fine-tuned to prevent demographic explosions. If, after a distance l, a b-fold splitting is introduced, the fraction of events starting from there is on average $be^{-l/\lambda}$, which becomes $b^2 e^{-2l/\lambda}$ after two splittings and so on. Therefore, we must select $be^{-l/\lambda} \sim 1 + \epsilon$, in order to get about $(1 + \epsilon)^n \sim 1 + n\epsilon$ walkers after n splittings. Splitting can also work in reverse. For instance, particles crossing the splitting surface at x_s coming from the deep layers, may receive a penalty, namely a weight reduction $w \leftarrow w/b$. Once its weight falls below a minimum treshold, w_{min}, the particle is eliminated. The splitting surface acts therefore like a two-way, non-symmetric, membrane.

7.7   Adaptive MonteCarlo

An alternative method to reveal particles in specific 'hot spots' is to bias the cross sections so as to enhance the survival probability around these regions. The procedure is as follows:

The probability to arrive at x and interact in the layer $[x, x+dx]$ is $P(x; dx) = \Sigma(x)e^{-\int_0^x \Sigma(y)dy} dx$. If we change Σ to Σ', a compensating weight w must be introduced, such that the above probability is left unchanged:

$$w\Sigma'(x)e^{-\int_0^x \Sigma'(y)dy} = \Sigma(x)e^{-\int_0^x \Sigma(y)dy}$$

This yields:

$$w(x) = \frac{\Sigma(x)}{\Sigma'(x)}e^{-\int_0^x [\Sigma'(y)-\Sigma(y)]dy}$$

For instance, by choosing $\Sigma' > \Sigma$, we give particles less chance to penetrate to large values of x, and viceversa in the opposite case. As a result, by fine-tuning Σ' we can focus attention either on the deep regions inside the domain or on the near-surface regions.

7.8   Fitness sampling

These are rather ruthless policies aimed at maximizing the global "fitness" of the population: at each given step one checks the actual weight of a given particle, and those which fall below a prescribed threshold w_{min} are eliminated with some probability. A suitable prescription is as follows: if $w < w_{min}$, compare $w\theta$, $\theta < 1$ being a tolerance parameter, with a uniform random number $0 < r < 1$. If $r > w\theta$ the particle is out, otherwise it is kept but with a reduced weight

$w\theta$. Low θ's goes with low tolerance. Zero-tolerance is hardly the best policy since it foregoes the benefits of 'redeemed' individuals, i.e. individuals whose fitness may grow significantly beyond average after a transient period of poor fitness. Further sophistication can be achieved by active control, i.e. by making the tolerance parameter θ a dynamic variable itself, responding, say, to variations of the overall fitness of the population.

7.9   Direct Simulation Monte Carlo

The examples discussed thus far refer to the *linear* version of the Boltzmann equation, describing situations in which test particles only interact with target atoms of the host media. Many applications, typically fluid-dynamics, but also classical and quantum kinetic theory, require the description of self-interactions between molecules. In this case, the (much more difficult) *non-linear* time-dependent Boltzmann equation must be solved. Symbolically, in Gain minus Losses form:

$$D_t f = Q(f, f) = G(f, f) - L(f, f) \tag{7.5}$$

where the collision term is given by the following quadratic integral:

$$G(f, f) = \int f(v') f(v1') P(v', v_1' \to v, v_1) dv_1 dv_1' dv' \tag{7.6}$$

and

$$L(f, f) = \int f(v) f(v_1) P(v, v_1 \to v', v_1') dv_1 dv' dv_1' \tag{7.7}$$

where the transition kernel P encodes the details of molecular interactions.

Owing to mass-momentum-energy conservation laws, the nine-dimensional integral reduces to a three-dimensional one, in the form:

$$Q(f, f) = \int [f(v) f(v_1) - f(v') f(v_1')] g\sigma(g, \Omega) dg d\Omega \tag{7.8}$$

where g is the magnitude of the relative speed $|v_2 - v_1|$ and Ω is the corresponding solid angle.

The non-linear Boltzmann equation is solved in two steps: 1) Streaming, 2) Collisions.

7.9.1   *Streaming*

The streaming term is dealt with by usual particle tracking schemes, the only condition being that particles should not travel more than a mean-free-path distance in a single time step. The computational grid (which serves only as a book-keeping device to collect physical observables such as gas density, speed and temperature) must be arranged accordingly, namely the grid spacing must be smaller than the particle mean free path. The right ordering of scales is therefore:

$$vdt < dx < \lambda$$

7.9.2   Collision

The collision term involves a very complicated 3-dimensional integral, which is best faced with Monte Carlo methods. Let us discuss this procedure in some more detail.

First, consistently with the short-ranged nature of the potential, we assume that only particles in the same spatial cell are allowed to interact. The expected number of collisions between particles i and j in a time interval dt is given by:

$$dN_{ij} = n g_{ij} \sigma(g_{ij}) dt$$

where n is the number density of the gas g_{ij} the magnitude of their relative speed and σ the corresponding cross section. This can be written in equivalent form as

$$dN_{ij} = dt/\tau_{ij}$$

which is de-facto a definition of the collisional time scale

$$\tau_{ij} = \frac{1}{n\sigma_{ij}g_{ij}}$$

In order to have a well-resolved simulation, we require

$$dt < \tau_{ij}$$

which means that the probability for a single pair to collide can be very small. The expected number of collisions in a given cell is the sum over all possible collisional pairs:

$$dN_{coll} = \sum_{i,j} dN_{ij}$$

The actual number of collisions to be performed at each time-step is the integer part

$$dI_{coll} = int(dN_{coll})$$

whereas the reminder $dR_{coll} = dN_{coll} - dI_{coll}$ is recorded and accumulated for subsequent time-steps. This evokes a sort of "horse-race" in which each cell undergoes its own numebr of collisions dI_{coll}. The cumulative number of collisions $I_{coll} = \sum_{timesteps} dI_{coll}$ can also be interpreted as a "biological time" (age) of the cell.

Having determined the number of collisions to be performed, we still have to decide which pairs should collide. This is done via usual hit/miss selection.

Once a pair i, j is selected at random within the cell, the corresponding collision probability is computed:

$$p_{ij} = dN_{ij}/dN_{coll}$$

which is indeed a real number in the range $[0, 1]$.

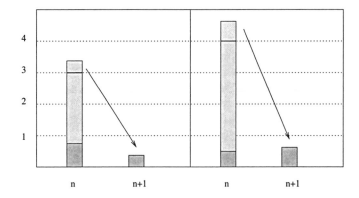

FIG. 7.2. *Horse race: the collision counter is incremented and the residual part is stored for the next time step.*

In actual practice, p_{ij} is distributed in a range $0 < p_{min} < p_{max} < 1$, and consequently the hit/miss test is most conveniently performed by drawing the random number in this range rather than $[0.1]$, so as to avoid needless rejects. It should be noted that computing p_{min} and p_{max} is computationally expensive, so that educated guesses and estimates are highly recommended whenever possible (a trivial, but unfortunately not very relevant case is provided by Maxwellian molecules, which have $\sigma g = const$ so that $p_{ij} = const$).

Since we make no distinction between coordinates of particles in the same cell, the collision matrix p_{ij} can be represented as a fully connected graph with N nodes and N^2 links. Due to the obvious properties $p_{ij} = p_{ji}$ and $p_{ii} = 0$ (a particle cannot collide with itself!), this graph only has only $N(N-1)/2$ distinct non zero links. The above expressions do include the possibility of particle recollisions: i.e a given particle can in principle collide with all its N potential partners. This violates the very spirit of one-body equations, and should not be permitted. In practice, once the collision (i^*, j^*) has taken place all links (i^*, j) and (i, j^*) are deleted, so that particles i^* and j^* are no longer available for recollisions. For each particle i, the choice of j can be made either randomly, or by selecting the most probable partner, $j_{max}(i)$, namely the one which maximizes p_{ij} for a given i.

7.9.3 *Kinematics of binary collisions*

Once a collisional event takes place, the particle speeds $\vec{v}_i, \vec{v}_j$ must be turned into the corresponding post-collisional values $\vec{v'}_i, \vec{v'}_j$. A convenient way to perform this task is as follows. Define the centermass speed as:

$$\vec{V} = \mu_1 \vec{v}_1 + \mu_2 \vec{v}_2$$

and peculiar speeds

$$\vec{c}_i = \vec{v}_i - \vec{V}$$

Mass, momentum and energy conservation imply that:

$$M' = M, \ \vec{V}' = \vec{V}, \ \vec{g}' = \vec{g}$$

where $M = m_1 + m_2$ is the total mass. From this, it follows that:

$$\vec{c}_1 = -\mu_2 \vec{g}, \ \vec{c}_2 = \mu_1 \vec{g}$$

where $\mu_i = m_i/M$ are the reduced masses. The only effect of collisions is to rotate the relative speed $\vec{g}$ by a given angle χ, which depends on the actual scattering potential via the conservation of angular momentum. The post collisional speeds are therefore computed as follows:

$$\vec{c}'_1 = -\mu_2 \vec{g}', \ \vec{c}'_2 = \mu_1 \vec{g}'$$

where $\vec{g}'$ is simply $\vec{g}$ rotated by an angle χ. The actual post-collisional speeds are then computed simply as:

$$\vec{v}'_i = \vec{V} + \vec{c}'_i, \ i = 1, 2$$

7.9.4 *Boundary conditions*

One of the main assets of KMC methods is the flexibility towards complex geometries, a flexibility which stems from the simplicity of particle trajectories. Boundary conditions are typically handled by monitoring whether the end-point of the particle path in the interval $[t, t + dt]$ is still within the computational domain, D or not. Assume for simplicity that the boundary ∂D is described by a single generalized coordinate $\xi = \xi_B$, such that $\xi < \xi_B$ characterizes the interior of domain D. The boundary test reads simply:

$$\xi[x(t + dt)] < \xi_B?$$

If the test fails, several actions may follow, depending on the type of boundary condition to be imposed.

A typical option is to reinject the particle with a speed drawn from a Maxwellian with the local wall temperature T_W and speed U_W:

$$v_n = -\sqrt{T_W} \ln r_n + U_{W,n}$$
$$v_t = -\sqrt{T_W} \ln r_t + U_{W,t}$$
$$\sin\theta = 2r_\theta - 1$$

where subscripts n, t stand for normal and tangential to the wall, respectively, and θ is the angle of the particle velocity to the normal to the wall.

With the speed set as indicated above, the particle is advanced within the gometrical domain over a time lapse $\delta t = dt - dt_W$, where $dt_W = |x_B - x|/|v|$ is the flight-time to the boundary.

The procedure discussed above is quite general and conceptually straightforward. In simple geometries, it involves at most the numerical evaluation of some elementary functions (squares, square roots). For complex geometries, however, fast methods to map the particle position x into the corresponding cell/element $C(x)$ of the computational grid, are needed. This is a non-trivial problem of computational geometry.

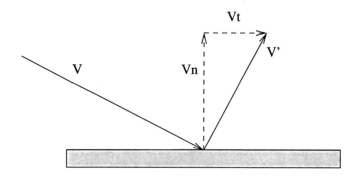

FIG. 7.3. *Wall collision: the new velocity of the particle is sampled from a given distribution. Note that the memory of the previous velocity is lost.*

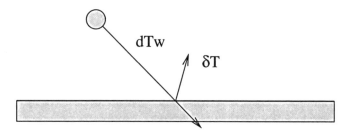

FIG. 7.4. *Then the particle is moved in space accordingly with the new velocity for the residual time δt.*

7.9.5 *Bird's method*

A fairly successfull Monte Carlo procedure to deal with this problem is due to G. Bird. Very schematically, the Bird's method consists of the following basic steps:

```
For each cell in configuration space:

With N=number of particles in the cell
up to Ncoll<N(N-1)/2 collisions are allowed

Clear the collision counter: icoll=0

1. Select a pair of particles (i,j) within the cell:

2. Compute:
     relative speed        : g=|vi-vj|,
     cross section         : sigma(g)
     collision probability : pcoll=N*g*sigma*dt/V
```

3. Throw a random number r in [0,1]

   If r>pcoll: Go to 1 (try again with different partners)

   Else   : Perform the mechanical collision:

   (v1,v2) --> (v1',v2')

     V=v1+v2, g=v2-v1, M=m1+m2

     g'=rotate(chi)*g: chi random in [0,2*pi]

     c1=-m2*g'/M, c2=+m1*g'/M

     v1'=V+c1, v2'=V+c2

4. Update counter and go on

   icoll=icoll+1

   If icoll<ncoll: go to 1

   Else: Move to next spatial cell

7.10   Random walks for PDE's

We conclude this Chapter with a mention to KMC as a procedure to solve partial differential equations on a grid. Consider for simplicity the simple advection-diffusion equation in 1D:

$$\partial_t P + U \partial_x P = D \partial_{xx} P$$

A simple discretization on a regular grid of spacing d and time-step h yields:

$$P(x,t) = aP(x-d, t-h) + bP(x, t-h) + cP(x+d, t-h)$$

where $a = Dh/d^2 - U/2d$, $b = 1 - 2Dh/d^2$ and $c = Dh/d^2 + Uh/2d$. Since $a+b+c=1$, and $a,b,c>0$ (iff $h < d^2/D + d/U$), the coefficients a,b,c can to all effects be regarded as probabilities per unit time to jump from one site to its neighbor, or eventually stay on the same site (b).

Therefore, a KMC procedure is immediately applicable:

For each lattice site:

```
1. Throw a random number r in [0,1]

2. Case 1:   0<r<a  : move right
        2:   a<r<a+b: don't move
        3:   a+b<r<1: move left
```

This is a noisy approximation to the continuum diffusion equation, with a slow convergence rate, hence rather inefficient for low-dimensional systems. However, it can be useful to describe complex diffusion phenomena with multiple concurrent processes, typically diffusion in solids and solid/fluid interfaces. It is also extremely useful in very high-dimensional systems such as those encountered in classical and quantum statistical mechanics.

7.11 References

1. P. Kevin Mac Keown, Stochastic Simulation in Physics, Springer Verlag, 1997.
2. G. Bird, Direct Simulation Monte Carlo, Oxford University Press, 1998.
3. F. Alexander, A. Garcia, The Direct Simulation Monte Carlo Method, Computers in Physics, vol. 11, n.6, p. 588, 1997.

7.12 Projects

1. Write a KMC transport code to compute the attenuation of a beam of gamma rays impinging on a slab of thickness $D = 30\lambda_A$. The only events are scattering (S) and absorption (A) with probability 0.7 and 0.3 respectively.

2. Write a DSMC code for the Rayleigh problem described below.
   A gas at temperature T_0 in a cubic box of heigth H, confined by an upper plate (reflecting wall) and a lower plate moving at speed U_w, both at temperature T_W. Parameter set-up: $kT_0 = 1$, $U_W = 0.1, kT_W = 1$, $m = 1, \sigma = 1, n = 0.001, \lambda = 225$ $H = 2250$, $dz = 0.2\lambda$ (mesh size) $dt = \tau/25$ (time-step), $Np = 50000$ (total number of particles) Run the simulation over 250 steps and compare the velocity profile with the analytical result provided in Ref. 3.

8

STOCHASTIC EQUATIONS

Stochastic equations play an important role in physics, engineering and biology. Generally speaking, stochastic equations arise from corresponding deterministic equations whenever a source of uncertainity (experimental uncertainity, noise due to fast degrees of freedom) needs to be taken into account. For most complex systems, this is the rule rather than the exception. In this Chapter we shall present a few simple methods for the numerical solution of Langevin equations, a special class of stochastic equations of great importance in physical applications.

8.1 Generalities

Let us consider the following stochastic Liouville equation:

$$\partial_t \psi = L_\omega \psi \tag{8.1}$$

where L_ω is the Liouville operator depending parametrically on the stochastic variable ω. The family of solutions generated by the different realizations of the random variable ω, define a stochastic process:

$$\psi_\omega(t) = e^{L_\omega t}\,\psi(0) \tag{8.2}$$

The expected value of a given physical observable $O(\psi)$ is therefore obtained by averaging over the realizations of the stochastic variable ω:

$$<O>(t) = \int O[\psi_\omega(t)]P(\omega)d\omega \tag{8.3}$$

where P is the probability distribution function (pdf) of ω. From the above expression, it is clear that, depending on the form of $P(\omega)$, the observable $O(t)$ can evolve in a way which looks very different (generally, smoother) from each of its individual realizations. The theory of stochastic differential/integral equations is a huge sector of modern mathematics, and we shall not deal with such a broad field here. We just want to emphasize that, in most practical cases, an explicit, analytical form of the solution $\psi_\omega(t)$ is not available, which is why numerical methods are required to compute the integral (8.3). In te following, we shall concentrate on a specific type of stochastic equation known as Langevin equations, which find a wide use in physics and all related disciplines.

8.2 The Langevin equation in statistiscal mechanics

As is well known, statistical mechanics provides three basic levels for the description of classical many-body systems:

- Fluid (Navier-Stokes)
- Kinetic (Liouville)
- Atomistic (Newton)

The lowest kinetic level (one-particle) associates with three fundamental equations:

- Vlasov equation (VE) (collisionless systems)
- Boltzmann equation (BE) (short-range collisions)
- Fokker-Planck equation (FPE) (long-range collisions)

The Langevin equation (LE) relates one-to-one with the Fokker-Planck equation, as we shall show shortly.

8.2.1   Langevin and Newton

Consider a system of N classical particles obeying Newton's equations:

$$\frac{dx_i}{dt} = p_i/m$$

$$\frac{dp_i}{dt} = f_i, \ i = 1, n$$

where f_i is the force acting on the i-th particle. Now split the particle coordinates and momenta into a coarse-grain and fluctuating components:

$$x_i = X_I + \xi_i, \quad p_i = P_I + \eta_i$$

where the coarse-grained index I runs from 1 to $N = n/B << n$, indicating that the coordinates (X_I, P_I) associate with a kind of macro-particle of mass $M = Bm$, collecting B microscopic degrees of freedom (B is sometimes called the 'blocking number'). Next, define a formal pair of coarse/fine-grain projector/coprojector operators P and $Q = 1 - P$ such that:

$$Px_i = X_I, \quad Qx_i = \xi_i, \quad P\xi_i = 0,$$

$$Pp_i = P_I, \quad Qp_i = \eta_i, \quad P\eta_i = 0$$

Upon coarse-graining the Newton equations with P, we obtain:

$$\frac{dX_I}{dt} = P_I/M$$

$$\frac{dP_I}{dt} = F_I + R_I$$

where $F_I = f(X_I)$ is the force computed at the mean-particle position, and the reminder

$$R_I = Pf(x_i) - f(Px_i)$$

collects the effects of the fast-degrees of freedom. Clearly, R_I is generally non-zero, unless the force $f(x)$ depends linearly on x.

The above set of stochastic equations is very similar to the Langevin equation, with R_I playing the role of noisy forcing.

A few analytical comments are in order.

Assuming the force $f(x)$ can be expanded in Taylor series, we obtain:

$$< f > \equiv Pf(x) = f(X) + \sum_{n=2} \frac{1}{n!} \frac{d^n f}{dx^n} (x = X) < \xi^n >$$

From this expression, we observe that fast degrees of freedom give rise to a hierarchy of effective forces, whose corresponding charges are given by the n-th order moment of the fast variables:

$$q_n = < \xi^n > / n!$$

Of course, in the case of linear forces this hierarchy simply vanishes. If the original force is an inverse power of the interparticle position $f(x) \sim x^{-\alpha}$, higher-order forces are increasingly more short-ranged, with exponent $-(\alpha + n)$. The 'strength' of the fluctuating degrees of freedom reflects into the decay-rate with n of the charge series q_n. Weak fluctuations yield fast-decaying series, strong fluctuations yield slowly-decaying series. For instance, assuming that fast degrees of freedom are distributed according to a zero-mean Gaussian with variance σ, we readily obtain that $q_n \sim \sigma^n$.

These considerations show the formal route from Newtonian dynamics to Langevin equations. Ideally, a renormalization-group treatment (recursive coarse-graining at various scales) would deliver the exact form of the entire hierarchy of moments. This task is clearly too difficult, and one should settle for less than that, namely low-order closures. This is precisely what the Langevin equation does: instead of deriving the residual term R_I ab-initio, suitable phenomenological expressions are postulated.

8.2.2 Langevin and Fokker-Planck

We have hinted at the fact that the Langevin equation is equivalent to the Fokker-Planck equation. Let us consider a particle, sometimes called the Rayleigh particle, which is subject to three type of forces: 1) external force $F(x)$, 2) dissipative force $F_d = -\gamma v$ expressing the particle interaction with the surrounding fluid (bath), 3) a random forcing R (we shall see below that R is not exactly a force). The particle velocity obeys the following equation of motion

$$m\dot{v} = F - \gamma v + R$$

In the absence of random forcing, the particle speed settles to the equilibrium value

$$v_{eq} = F/\gamma,$$

The random term R allows random excursions around this equilibrium value.

We assume the random forcing corresponds to a zero-mean delta-correlated (white) noise with variance σ^2:

$$< R >= 0$$
$$< R(t)R(t') >= \sigma^2 \delta(t - t') \tag{8.4}$$

where $< . >$ stands for time-average (note that it is σ, and not R, which has dimensions of a force!).

Now we introduce the distribution function $p(v, t)$ expressing the probability of finding a particle at time t with speed v, and look for the kinetic equation whose underlying microdynamics is given by the Langevin equation. The distribution $p(v, t)$ fulfills a master equation of the form:

$$p(v, t) = ap(v - dv, t - dt) + bp(v, t - dt) + cp(v + dv, t - dt)$$

where a, b, c are (dimensionless) transition probabilities fulfilling the normalization condition $a + b + c = 1$. Expanding to the first order in dt and second order in dv, we readily obtain:

$$\partial_t p = (c - a)\frac{dv}{dt}\partial_v p + (c + a)\frac{dv^2}{dt}\partial_{vv} p$$

The case $a = 1$, $b = c = 0$ is easily recognized to correspond to the deterministic motion $dv = Fdt/m$. However, due to the presence of the stochastic term R, we cannot write $m\dot{v} = F + R$ because, due to the stochastic component R, the particle does not move along a smooth, differentiable, trajectory.

As a result, one must take the average form of the above equation. This reads:

$$\partial_t p + V\partial_v p = D\partial_{vv} p \tag{8.5}$$

This is indeed a Fokker-Planck equation, with drift and diffusion coefficients defined as

$$V = (c - a)\lim_{dt \to 0}\frac{< dv >}{dt}$$

and

$$D = (c + a)\lim_{dt \to 0}\frac{< dv^2 >}{dt}$$

How do we recouncile this picture with the Langevin equation?

The correct viewpoint consists in recasting the Langevin equation into the so-called *Ito* differential form:

$$mdv = (F - \gamma v)dt + Rdt^{1/2}$$

It is immediately seen that this reflects the finitess of the variance σ of the stochastic process R. Write:

$$m < dv >= (F - \gamma < v >)dt + < R > dt^{1/2} = (F - \gamma < v >)dt$$

where brackets stand for average over the noise, and we have used the fact that, by definition, R has zero-mean. As a result, the mean speed obeys a deterministic equation

$$m\frac{d<v>}{dt} = F - \gamma <v>$$

However, since v is a stochastic process, higher order moments -and most notably the variance-need also to be computed.

Squaring the Ito-form of the Langevin equation, we obtain:

$$m^2 < dvdv >= (F - \gamma <v>)^2 dt^2 + <RR> dt$$

As a result, in the limit $dt \to 0$, we obtain:

$$D \equiv lim_{dt\to 0}\frac{<dvdv>}{dt} = \sigma^2/m^2$$

where

$$D = \gamma\frac{k_B T}{m}$$

is the diffusion coefficient in velocity space.

Since the Dirac delta has dimension of an inverse time, we identify the variance of the random forcing as follows:

$$\sigma^2 = mk_B T\gamma dt$$

Since the Langevin equation is equivalent to the Fokker-Planck equation, we may wonder why should we use the former instead of the latter?

The advantage, as usual, becomes compelling for high-dimensional problems, where a grid-based representation of the probability distribution $p(v, t)$ becomes unviable. For these problems, the Langevin equation provides a naturally adaptive grid-free method to solve the Fokker-Planck equation. In passing, we wish to add that is even more true in statistical and quantum field theory, where one is confronted with virtually infinite-dimensional (functional) spaces. The Parisi-Wu stochastic quantization method is a well-known example in point.

8.3 Brownian motion and anomalous diffusion

The most typical stochastic process is the Brownian motion of a material particle within, say, a (very viscous) fluid flow. The Ito formulation of Brownian motion is:

$$dx = vdt + R\sqrt{dt} \tag{8.6}$$
$$dv = adt$$

with the usual properties:

$$<R>= 0 \tag{8.7}$$
$$<R(t)R(t')>= \sigma^2\delta(t' - t)$$

where the variance is related to the diffusion coefficient in configuration space:

$$D = lim_{dt \to 0} \frac{<dxdx>}{dt} = k_B T/m\gamma$$

It should be noted that Brownian motion is less regular than the motion of the Rayleigh particle. In fact, the latter has $dv \sim dt^{1/2}$, namely $dx \sim dt^{3/2}$, which means that the velocity is a continuous but not differentaible process, whereas the particle position is both continuus and diferentiable. Brownian motion, on the contrary, is not differentiable, since $dx \sim dt^{1/2}$.

Brownian motion is a specific instance of a more general class of stochastic processes known as *fractional random walks*. Non-linear physics presents us with many istances of *fractional* random walks, with a varying degree of roughness. Mathematically, the condition:

$$D_a \equiv lim_{dt \to 0} <dx^a> /dt$$

defines a random walk with scaling exponent $1/a$. The larger is a, the roughest is the process.

Special cases are:

- Advection $(a = 1)$
- Hyper-Diffusion $(1 < a < 2)$:
  Typically observed in turbulence, where coherent blobs of fluid can engulf the particle and speed it up via a 'go-with-the-flow' effect.
- Normal Diffusion $(a = 2)$
  The standard random walk.
- Hypo-Diffusion $(a > 2)$
  These are random walks in heterogeneus media with traps and cages which result in slowing down of the particle motion.

In principle, these fractional randow walks associate with a corresponding fractional Fokker-Planck equation. Hyperdiffusion is associated with "wilder" stochastic processes than Brownian motion, known as Levy processes. A Levy process of degree k is defined by the condition that only moments up to order k are finite. For instance, a process with probability distribution

$$p(x) = \frac{1}{1 + x^2}, \quad -\infty < x < \infty$$

has infinite mean, hence $k < 1$. The numerical simulation of these phenomena with partial differential equations requires non-trivial extensions of standard calculus, known as fractal calculus. We shall not deal with this still-developing subject here.

We proceed instead to discuss the numerical integration of the Langevin equation in Ito form.

8.4   Numerical integration of stochastic equations

The numerical integration of stochastic differential equations (SDE's)is a rather difficult topic which involves a number of mathematical subtleties.

Here we only sketch briefly some major point.

We restrict our attention to Langevin-like equations of the form:

$$dX_t = a(X,t)dt + b(X,t)dW_t$$

where X_t is a stochastic process, a the deterministic interaction and dW_t is a differential Wiener process. The above stochastic differential can be recast into an equivalent integral form

$$X_t = X_0 + \int_0^t a(X_u, u)du + \int_0^t b(X_u, u)dW_u$$

Most numerical integration schemes for SDE's are based on stochastic versions of Taylor expansions of the above integral expressions. The problem, of course, is to give proper meaning to the stochastic differential dW_t and the corresponding stochastic integral. To be noted the fact that the stochastic integral does *not* fall within the standard categories of Riemann, Lebesgue or Stjelties integral, essentially because the result of the associated discrete summation in the limit $dt \to 0$ does depend on the specific discrete sequence t_n. Roughly speaking this is again because the Wiener process is a Brownian motion whose trajectory exhibits a diverging arc-length as the time-step goes to zero (distinctive feature of fractal objects).

Two major choices are the Ito and Stratonovich rules.

Ito-calculus:

$$\Delta X = b(X_0, 0)[W(h) - W(0)]$$

Stratonovich-calculus:

$$\Delta X = \frac{1}{2}[b(X_0, 0) + b(X_{h/2})][W(h) - W(0)]$$

where we have set $h \equiv dt$.

The Ito-version of stochastic difference equations is simpler, hence most widely used. The heuristic recipe for a standard Wiener process in the infinitesimal interval $[0, h]$, is to take:

$$dW \sim W(h) - W(0) \sim rh^{1/2}$$

where r is a random number in $[0, 1]$, with zero mean and unit variance. This expression assumes that the Wiener process does possess a finite variance, so that the limit $(dW)^2/h$ stays finite as h is sent to zero (roughly speaking, the Wiener process has finite derivative of order $1/2$). As anticipated, this is not true in the case of Levy processes. The simplest numerical schemes associate with Ito calculus are the Euler and Heun schemes.

8.4.1 *Euler scheme*

The Euler scheme is the direct Ito-analogue of standard Euler time marching for deterministic equations. It reads as follows:

- $X_h = X_0 + a_0 h + b_0 w$

where $w \equiv \Delta W \sim r h^{1/2}$.

The Euler scheme is Ito $S0.5$, i.e. it offers 0.5-order accuracy in strong-sense, strong meaning point-like convergence to the exact result for each value of t, as h goes to zero. This is the simplest and most popular form in physical applications. However, just like the Euler scheme for deterministic equations, this scheme has poor accuracy. A better method is the Heun scheme.

8.4.2 *Heun scheme*

The Heun scheme is the direct Ito-analogue of standard improved-Euler (trapezoidal rule) time marching for deterministic equations.

It reads as a two-step predictor-corrector scheme:

- Predictor: $\tilde{X}_h = X_0 + a_0 h + b_0 w$
- Corrector: $X_h = X_0 + \frac{h}{2}[a_0 + \tilde{a}_h] + \frac{w}{2}[b_0 + \tilde{b}_h]$

The Heun-scheme is Ito $W2.0$, i.e. second order weakly convergent, where weak convergence means convergence in L_2 norm.

8.5 Stochastic integration of the Langevin equations

Langevin equations for a system of N classical particles associate with a $6N$-dimensional stochastic process for the particle positions x and particle speed v (particle index relaxed for simplicity):

$$dx = vdt \tag{8.8}$$
$$dv = adt + dW \tag{8.9}$$

For instance, Euler marching reads as follows:

- $x(t + dt) = x(t) + v(t)dt$
- $v(t + dt) = v(t) + a(x, t)dt + r\sqrt{dt}$

where r is random number in $[0, 1]$ with the desired variance.

8.6 Applications

The use of Langevin equations spreads across a huge variety of physical applications. Here we briefly illustrate some representative examples.

8.6.1 *Stochastic resonance*

Stochastic resonance (SR) is the physical process by which external or internal noise operates on a non-linear multistable system, modulated by a weak periodic load, to induce, or enhance, switching events among the stable states. Stochastic resonance arises from the interplay between the modulation frequency and Kramers escape rate from stable states. A typical SR system is a particle in the double-well potential with small-amplitude low-frequency oscillating load and additive noise:

$$\ddot{x} = ax - bx^3 + A\cos(\omega t) + \xi$$

Low amplitude means that A is always smaller than the barrier heigth $H = a^2/4b$ so that, in the absence of noise, the particle cannot exit the stable state she is trapped in. When noise is on, there always is a non-vanishing probability for the particle to cross the barrier, of the order of $exp(-H/k_BT)$, where T goes with the variance of the noise. If $k_BT << H$ the particle virtually never escapes, so that the switching rate between the stable states is zero. In the opposite limit, the particle escapes all the time but in a totally random and incoherent fashion. The interesting thing is that there exists an intermediate optimal value of T, such that the escape rate responds in fine-tune with the external modulation. In other words, noise helps driving a coherent signal!

Numerical simulation is very valuable because the stochastic process is non-stationary, so that analytical techniques have little horizon. This is even more so for the case of coupled SR oscillators of the form:

$$\ddot{x}_i = ax_i - bx_i^3 + A\cos(\omega t) + \xi_i + \kappa(x_{i+1} - 2x_i + x_{i-1})$$

where spatial inhomogeneity couples with the previous mechanisms.

8.6.2 Dynamic phase-transitions: the Landau-Ginzburg equation

Langevin equations are not restricted to particle motions, they apply to classical (and quantum) field theory as well. A paradigmatic application is the study of phase-separation in binary fluids. Denoting by ϕ the ratio of light/heavy fluid (the so-called) phase, the dynamics of the binary fluid is governed by the following Landau-Ginzburg equation for the order parameter:

$$\partial_t \phi = \Delta\phi - F(\phi) + \xi$$

where ξ is a zero-mean delta-correlated (in time and space) noise:

$$< \xi(x,t)\xi(y,u) > = \sigma^2\delta(x-y)\delta(t-u)$$

and $\sigma^2 = k_BT/m$. Upon using a simple centered-finite difference scheme, we obtain:

$$\phi(l,n+1) = (1-2d)\phi(l,n) + d(\phi(l+1,n) + \phi(l-1,n)) + F[\phi(l,n)]h + R(l,n)$$

where l and n denote discrete time and space respectively, and we have set $h \equiv \Delta t$ and $d = Dh/\Delta x^2$. The random term is constructed as follows:

$$R(l,n) = r\frac{\sigma^2}{h^{1/2}\Delta x^{1/2}}$$

where r is a uniform random number in $[0,1]$.

To be noted that the square roots of the space-time mesh in the denominator of the above expression are due to the discrete representation of the Dirac delta: $\delta(x - y) \sim \frac{\delta_{ll'}}{\Delta x}$ and $\delta(t - u) \sim \frac{\delta_{nn'}}{h}$.

It is clear that due to this square root dependence on the denominator, the convergence of the numerical scheme in the limit of vanishing time-step and space-meshing is a very delicate issue.

8.6.3 Interface growth: The Kardar-Parisi-Zhang equation

Many non-linear phenomena of irreversible growth, typically interfaces, are well described by the famous Kardar-Parisi-Zhang (KPZ) equation:

$$\partial_t h = \nu \partial_{xx} h + \frac{\lambda}{2}(\partial_x h)^2 + \xi \qquad (8.10)$$

where $h(x, t)$ is the heigth of the interface at position x at time t. The first term at the right hand side represents relaxation of the interface due to surface tension ν. The second term represents the lateral growth of the surface (in the direction non-normal to the interface) and finally ξ is a random space-time delta-correlated noise. The KPZ equation has two remarkable mappings which greatly facilitate analysis of its scaling properties.

The transformation:

$$lnV = \frac{\lambda}{2\nu} h$$

yields:

$$\partial_t V = \nu \partial_{xx} V + \frac{\lambda}{2\nu} \xi V \qquad (8.11)$$

This is a diffusion equation in a random potential (such as Anderson's localization). The transformation:

$$u = -\partial_x h$$

yields:

$$\partial_t u + \lambda u \partial_x u = \nu \partial_{xx} u - \partial_x \xi \qquad (8.12)$$

This is (for $\lambda = 1$) the random Burgers equation for a pressure-free fluid.

The main theoretical question is to identify the dynamic scaling exponents associated to the interface growth dynamics:

$$w(x, t) \sim L^\chi w_0(t/L^z)$$

where w is the width of the interface (rms of the variance $< (h - \bar{h})^2 >$, and L is the linear size of the domain.

Remarkable special cases of KPZ are:

- $\nu = \lambda = 0$: Random deposition model
  The interface grows like $t^{1/2}$.
- $\lambda = 0$: Ideal (linear) interface
  We have $z = 2$ and $\chi = (2 - d)/2$ in dimension d.

- $\lambda \neq 0$: Non-linear growth
  We have $z = 2/3$ and $\chi = 1/2$ in dimension d.

The KPZ is yet another form of non-linear Langevin equationi, and it can be numerically integrated as discussed previously. A simple centered difference scheme combined with Euler time marching, yields:

$$h(l, n+1) = (1 - 2N)h(l, n) + 2N[h(l+1, n) + h(l-1, n)]$$
$$+ \Lambda[(h(l+1, n) - h(l-1, n))^2] + R(l, n)$$

where $N \equiv \nu dt/dx^2$, $\Lambda \equiv \lambda dt/dx^2$ and the random term is treated as in the previous section.

8.6.4 Stochastic quantization

We mentioned that Langevin-type equations are very useful in infinite-dimensional frameworks, such as statistical field theory and quantum field theory.

Indeed, it is known that the N-point correlation function of a quantum bosonic field is given by:

$$S_N(x_1, x_2, \ldots x_N) = \int \prod_i^N \phi(x_i) d\mu[\phi]$$

where $d\mu$ is a probability measure on a functional space, given by

$$d\mu[\phi] = e^{S[\phi]} d\phi$$

and $S[\phi]$ is the classical Euclidean action of the field.

As shown by Parisi and Wu, the theory can also be formulated as a stochastic functional differential equation by introducing a fictitious time t (five-dimensional time):

$$\partial_t \phi = F(\phi) + \eta(x, t) \tag{8.13}$$

where η is a white noise and the 'force' F is given by the functional derivative

$$F(\phi) = -\frac{\delta S}{\delta \phi} \tag{8.14}$$

By integrating the functional Langevin equation in time, quantum expectation values, such as $S_N(x_1 \ldots x_N)$, can be computed by replacing ensemble averages with time averages (ergodicity assumption):

$$S(x_1 \ldots x_N) \sim lim_{T \to \infty} \frac{1}{T} \int_0^T \prod_i^N \phi(x_i, t) dt \tag{8.15}$$

where T is the time span of the simulation.

The numerical integration of the functional Langevin equation (8.14) cannot be performed using a grid method because of the extremely high-diemnsionality of the problem (virtually infinite). However, particle methods similar to those

used in the Quantum Monte Carlo are easily adapted to the functional Langevin
equation as well. First, one replaces the continuum, infinite-dimensional, field
$\phi(x,t)$ with a discrete set of, say, N discrete real variables $\Phi_N \equiv \{\phi_j(t),\ j = 1, N\}$. Subsequently, each realization of the vector Φ_N is represented as a walker
$\phi_i,\ i = 1, NW$. Each walker ϕ_i is advanced according to discretized Langevin
equation:

$$\phi_i(t + dt) = \phi_i(t) + F_i(\phi_1 \ldots \phi_{NW})dt + R_i$$

As usual, the expectation is that the Langevin dynamics serves as a naturally
adaptive numerical method, sending the walkers only in the important regions
of the N-dimensional discretized functional space.

8.7   References

1. P. E. Kloeden, E. Platen, Numerical solution of stochastic differential equations, Springer Verlag, 1996
2. K. Elder, Langevin simulation of nonequilibrium phenomena, Computers in Physics, vol. 7, n. 1, p. 27, 1993.

8.8   Projects

1. Solve the Langevin equation associated with the Landau-Ginzburg equation discussed in this chapter (for the details, see Ref. 2).

8.9   Sample program

```
c  Langevin method for Fokker-Planck: dP/dt=div(FP+DdP/dx)
c  should be more efficient than pure MonteCarlo
c  because each move is accepted
c  ==========================================================
           parameter(NX=128,NB=NX,MT=100000)
           dimension eta(MT)
           dimension ncount(0:NB)
c  ----------------------------------
           write(6,*) 'n.of walkers?'
           read (5,*)  NW
           write(6,*) 'n.of steps?'
           read (5,*)  NT
           write(6,*) 'dif and drift'
           read (5,*)  dif,udrift
           write(6,*) 'dt'
           read (5,*)  dt

       dx=1.

       do i=0,NB
```

```
                ncount(i)=0
                end do

                iseed = 3452987
                do i=1,NW
                xold = NX/2
c NT gaussian random numbers with variance dif
                var = sqrt(2.*dif)
                do it=1,NT,2
                  call rgauss(iseed,eta(it),eta(it+1),var)
                end do
                do it=1,NT
                 xnew = xold+udrift*dt+eta(it)*sqrt(dt)
                 if(xnew.gt.NX) xnew=xnew-NX
                 if(xnew.lt.0)  xnew=xnew+NX

                 write(36,*) it,xnew

                 ib   = xnew/dx
                 ncount(ib)=ncount(ib)+1

                 xold = xnew
                end do
c end of a given walker
                end do
c ------------------------------------
                do ib=0,NB
                 write(61,*) ib,ncount(ib)/NW
                end do

                stop
                end

c ===================================
          subroutine rgauss(iseed,rg1,rg2,var)
c ===================================
c sample from Gaussian
c -----------------------------------
          pi = 4.0*atan(1.0)

          ri = ranpang(iseed)
          r1 = -alog(1.0-ri)
          r1 = var*sqrt(2.0*r1)
```

```
ri = ranpang(iseed)
r2 = 2.*pi*ri
rg1= r1*cos(r2)
rg2= r1*sin(r2)

return
end
```

Elenco dei volumi della collana
"Appunti"
pubblicati dall'Anno Accademico 1994/95

SAURO SUCCI, *An Introduction to Computational Physics. Part I: Grid Methods,* 2002

DORIN BUCUR, GIUSEPPE BUTTAZZO, *Variational Methods in Some Shape Optimization Problems,* 2002

EDOARDO VESENTINI, *Introduction to continuous semigroups,* 2002

ANNA MINGUZZI, MARIO TOSI, *Introduction to the Theory of Many-Body Systems,* 2002

SAURO SUCCI, *An Introduction to Computational Physics. Part II: Particle Methods,* 2003

Fotocomposizione "CompoMat" Loc. Braccone, 02040 Configni (RI), Italy
Finito di stampare per conto della "CompoMat" dalla Nuova Grafica 86 nel maggio 2003